A Student Guide to AI and Robotics

This book provides a practical and accessible introduction to AI and robotics, presenting the foundational technologies driving the development of intelligent systems. It covers key topics such as machine learning, perception, control, and autonomous decision-making, balancing theoretical concepts with practical applications.

Through selected examples and case studies, the book explores robotics platforms and applied AI technologies used across industries, offering insights into their transformative potential in real-world scenarios. With a focus on hands-on learning, it introduces programming languages and tools commonly used in AI and robotics, empowering readers to apply concepts and techniques effectively. It also addresses the importance of building secure, responsible systems and examines the future of human-AI collaboration.

An essential resource for introductory AI and robotics courses, this book provides the foundational knowledge needed to advance in these dynamic fields.

A Student Guide to AI and Robotics

With Real-World Case Studies and Programming Examples

Mahwish Zahara,
Tamour Raza, and
Obianuju Blossom Ochuba

CRC Press
Taylor & Francis Group
Boca Raton London New York

CRC Press is an imprint of the
Taylor & Francis Group, an **informa** business

Designed cover image: Shutterstock

First edition published 2027
by CRC Press
2385 NW Executive Center Drive, Suite 320, Boca Raton FL 33431

and by CRC Press
4 Park Square, Milton Park, Abingdon, Oxon, OX14 4RN

CRC Press is an imprint of Taylor & Francis Group, LLC

ISBN: 978-1-041-28261-7 (hbk)
ISBN: 978-1-041-28256-3 (pbk)
ISBN: 978-1-003-75850-1 (ebk)

DOI: 10.1201/9781003758501

Typeset in Times
by KnowledgeWorks Global Ltd.

Contents

Preface

Artificial intelligence (AI) and robotics are technologies that surround us every day. You likely interact with AI-powered virtual assistants, recommendation systems, smart homes, self-driving cars, and service robots daily. AI and robotics will change how we live, work, play, and think, and you can be part of that change. But to do so, you need to learn more than theory. Students of AI and robotics need opportunities to see these technologies in action.

A Student Guide to AI and Robotics is your opportunity to learn more about how robots and intelligent systems are designed, developed, programmed, and used in the real world. This book is an introduction to AI and robotics for students of any background who want to learn about how smart machines work. This book is not just about theory. It uses examples and case studies to provide a practical introduction to AI and robotics. Topics covered include programming concepts, mathematics, embodied intelligence, perception, learning, planning, and human-robot interaction. In addition to technical topics, we include case studies and examples of real-world applications of AI and robotics, from self-driving cars and drones to robots in factories and warehouses, as well as ethical considerations and new advances in AI. Through this book, you will gain a better understanding of the technology you see around you and begin to think like a roboticist.

Robotics is inherently interdisciplinary. Designing, developing, and using intelligent systems requires knowledge of computer science, engineering, and mathematics. It also requires students to understand ethics, social issues, and career skills. The technical topics covered in the early chapters of this book will be foundational for understanding more advanced material later. In later chapters, we discuss important professional topics that every student should know, including security, ethics, and project management. We also discuss the future of AI and humans working together.

In addition to offering theory and real-world examples, we also provide programming examples throughout the book. While not every student is a programmer, familiarity with programming concepts and techniques is a hugely valuable skill. Programming examples are included throughout the book, where applicable. Furthermore, we try to use programming languages and robotics platforms that are widely used in both industry and academia, such as Python and the Robot Operating System (ROS).

This book is for students. If you are a student learning about AI and robotics for the first time, this book will help you learn the foundational skills and knowledge you need to further your education. If you are already studying AI and robotics, this book can help you apply what you know. By the end of this book, we hope that you will have a deeper understanding of AI and robotics. You will learn how to think like a roboticist.

About the Authors

Mahwish Zahara is an artificial intelligence (AI) lecturer, published author, and guest speaker, specialising in marketing strategy, data science, and applied AI. She designed and launched one of the UK's first ever Master's courses in AI in Marketing, which enabled students to learn machine learning workflows, advanced data applications, and how to apply them to real world business problems.

An expert in marketing strategy with a PhD in Business and Marketing, her research explores the intersection of digital technologies and marketing, with a focus on generative AI, marketing automation, consumer insights, and algorithmic decision-making. Through her research and teaching, she embeds new technologies into the learning journey to develop technical and critical thinking skills for the AI-enabled future global economy.

She has contributed curriculum design and innovation at academic and commercial levels, developing frameworks for ethical and strategic use of large language models for teaching and research outputs. Delivering hands-on experience and an industry-facing approach to learning, her students gain experience with AI model programming workflows, large model architectures, and cloud-based computing platforms while developing critical thought around their limitations, metrics for performance, and real-world applicability.

She investigates frontier technologies like quantum computing, AI-driven robotics, and the next generation of intelligent systems. With recent publications and editing AI in marketing texts on AI-powered customer experience, conversational AI, and ethics of AI in marketing, Dr Zahara continues to contribute to innovation at the forefront of AI and marketing.

Tamour Raza conducts research around artificial intelligence and robotics engineering, specialising in machine learning, algorithmic decision-making, and designing intelligent socio-technical systems. His work has covered both theoretical and computational principles for supervised and unsupervised learning; prediction; optimisation-based decision-making; and modelling systems that learn through feedback commonly used in robotics and autonomous systems.

He is particularly interested in how intelligent systems behave at scale, deployed in intricate settings defined by uncertainty, variability, real-time operations, and large, heterogeneous end-users. Tamour explores system performance, robustness, and adaptation, while considering broader implications such systems have when deployed at scale.

Tamour's previous academic research has surrounded artificial intelligence systems within industry sectors, publishing papers around algorithmic systems in marketing and how technologies such as recommender systems, customer intelligence models, and real-time decision engines affect organisational decision-making processes and end-user experiences. He is the co-author of AI textbook *A Student Guide to AI and Robotics*, which closely ties his research to real-world applications.

Tamour takes a socio-technical approach to his research and considers, alongside raw technical performance metrics of machine learning systems, how they may be used at an organisational level, and what their broader ethical and societal consequences are. He aims to design systems that are rigorous, interpretable, and accountable.

Obianuju Blossom Ochuba specialises in robotics and artificial intelligence with a focus on intelligent systems, automation, and data-driven solutions. She graduated with a Master of Science degree in Robotics and Artificial Intelligence as well as a degree in Electrical and Electronic Engineering.

She has experience in robotic perception, adaptive control, automation, and intelligent data systems. She also possesses knowledge in programming and algorithm development, which she has utilised while working in technical and analytics roles in the industry. She has worked on projects involving automation pipelines, data systems/architecture, and AI-driven decision-making systems.

Her expertise includes building and testing smart systems in real-world scenarios. She's always curious about how well these systems perform and how they can be scaled. Obianuju often implements her work from an engineering standpoint and an applied standpoint, as she loves to understand the theory behind practical implementations. She's passionate about creating and helping to build robust and scalable AI systems as well as helping technology function in complex real-world scenarios.

Part I

Foundations of Intelligent Systems

1 Understanding Embodied Intelligence

1.1 INTRODUCTION

Concepts of embodied intelligence have historically stood in opposition to disembodied artificial intelligence (AI), which describes computational systems operating on abstract representations of reality, such as information processing systems running in a digital environment. This contrasts with the recent paradigm shift of AI functioning on computers physically embodied in robotic agents that perceive their environment, reason about their perception, and act on those reasons to affect change. While many of these agents operate in cyber-physical environments that are difficult to distinguish from cyberspace itself, embodied intelligence machines have sensory and actuation interfaces that persistently sense their environment, respond to their perceptions, and act to change their environment in feedback loops.

Operating in physical reality has replaced computation on raw data as the key differentiation between traditional AI and embodied intelligence. Closed-loop perception, control, and actuation have furthered popular notions of embodied intelligence to include coupling between sensing and acting as intelligent behaviour arises from these dependencies. A robot's action changes its sensor measurements and drives its learning. This requires robots to interpret sensory information from the world, reason about what actions to take in the world, and act by triggering effectors that can change the world. Robots may use world models to inform their actions and intelligently navigate uncertainty through mechanisms like neuromorphic or bioinspired controls. The same coupling does not exist for many disembodied AI use cases, such as image classifiers trained on photographs or game players instructing simulated robots to complete tasks. These systems tend to be static and operate on their inputs at their own pace.

Embodiment matters because intelligence without embodiment cannot physically exist in the world. They must reason in real time under uncertainty resulting from noisy sensory information, partial information, and non-stationary environments. Perception, controllers, and effectors are all hardware systems that will never operate in the real world without failures, regardless of how sophisticated the algorithms are themselves. Intelligence without embodiment also does not need to consider physical realities such as sensing, actuating, and the environmental coupling between the two. Robots need sensors to perceive the world around them, controllers and planners to decide how to act, and actuators (or soft bodies) to move or provide that environmental coupling. Analysing how these components work together to enable intelligent behaviour in robots and other embodied agents will help advance the fields of healthcare robotics, robotics in manufacturing, soft robot

DOI: 10.1201/9781003758501-2

exploration, and more. Understanding embodied intelligence helps us ask the right questions about how intelligence operates outside of a simulator.

1.2 FOUNDATIONS OF EMBODIED INTELLIGENCE

In embodied intelligence, intelligence is achieved through close coupling of an agent's body with its environment, such that action, perception, and learning are formed through sensorimotor interactions with the environment (Liu et al., 2024; Liu et al., 2025). Embodied intelligence requires sensing, processing, and acting tightly coupled through feedback, commonly implemented as closed-loop control systems that coordinate perception, planning, and action at runtime (Mengaldo et al., 2022). Intelligence arises not just from algorithms alone but depends on an embodiment's hardware and software, including its morphology, materials, sensors, actuators, and computation that together enable adaptive behaviours (Wang et al., 2025).

Embodied intelligence typically consists of three key ingredients: sensors, actuators, and control that realise perception–action loops (Yang et al., 2024). Sensors allow an embodied system to sense its environment by measuring some quantity in the world, such as distance, light intensity, motion, or temperature, often through flexible multimodal sensors capable of high-resolution tactile sensing that passively acquire information about their environments (Zhao et al., 2024). Actuators enable embodied systems to act through movement or mechanical work on their environment, spanning from rigid motors to increasingly soft and bioinspired actuators for greater compliance (Li et al., 2025). Control systems provide decision-making capabilities that interpret sensory input and decide how to act based on that information using various algorithms, including reinforcement learning, model-predictive control, and neuromorphic control architectures (Radosavovic et al., 2023).

Systems can leverage all three of these ingredients in tandem. For instance, embodied systems often exhibit locomotion capabilities, where a robot navigates through an environment. The robot may use sensors to perceive obstacles via vision, touch, or another modality, then send this sensory data through a controller which can process that information to select an appropriate motor action to steer the robot away from the obstacle, using feedback to correct for deviations and maintain the course (Liu et al., 2024; Radosavovic et al., 2023). Such control can be done online and reactively so that robots can dynamically adapt to changes in their environments. Many methods for enabling soft robots and legged robots to handle terrain variations, disturbances, and uncertainty rely on closed-loop sensorimotor control systems (Li et al., 2025).

Another important feature of embodied intelligence is feedback. Because an agent's actions affect its environment, and in turn these changes are sensed by the agent, embodied intelligence gives rise to recursive loops of perception and action. These feedback loops allow embodied systems to adapt to their environment (Liu et al., 2025). Feedback allows embodied intelligence to enable new behaviours that are better and more adaptive. These include self-adaptive intelligent manipulation, environment-aware locomotion, task-level autonomy in unstructured environments, etc. (Li et al., 2025; Wang et al., 2024). Because of this, embodied intelligence can deal with many real-world environments where there are uncertainty and unpredictability, whereas traditional AI applications are limited to passively processing datasets (Liu et al., 2024; Wang et al., 2025). Embodied intelligence extends AI into the physical world.

1.3 MODELLING INTELLIGENT BEHAVIOUR IN ROBOTICS

Building intelligent behaviour into robotic systems can often be organised using a pipeline that maps perception to action decisions. Recent surveys on embodied robotics and autonomous robotics both emphasise the importance of having a pipeline to model perception, decision-making, and control behaviour (Liu et al., 2024). In this chapter, we will use an augmented framework that Sense–Think–Act paradigm as a base definition to analyse embodied system behaviours operating in real-world environments with explicit steps for perception, cognition, control, and continuous adaptation ability (Ren & Kim, 2025).

The Sense–Think–Act–Learn (STAL) architecture framework characterizes intelligent behaviour as a continuous, recursive cycle, where an agent senses the environment around itself, reasons about what it senses, acts on the world, and finally learns from the result. It is an extension to the traditional Sense–Think–Act architecture typically used for robotic behaviours, with the addition of learning(Ren & Kim, 2025). Sense is the process of observing the states of the environment through sensory inputs such as cameras, range sensors, or inertial measurement units (IMU) that provide vision, geometric, and motion cues to higher-level cognitive processing pipelines (Ren & Kim, 2025). State observation generally encodes robot status and contextual information to be used by other processing stages. Think describes computational steps that take as input sensed information and processes it using algorithmic/logical control rules or learned machine intelligence, such as deep perception networks, planning algorithms or reinforcement learning policies (Thaker, 2024). Act corresponds to the actuation of robotic movements and hardware that translate computational decisions into changes in the real world, expressed through robotic manipulators, wheels, or specialised movement hardware (Ren & Kim, 2025). Learn augments the conventional framework to include continuous closed-loop adaptation of behaviour through environmental feedback and experience, using methods such as embodied reinforcement learning, imitation learning, or model-based self-learning with digital twins in both simulated and physical worlds (Liu et al., 2025).

Figure 1.1 depicts the STAL framework as a feedback loop rather than a linear model. The focus is on the ongoing feedback between environment and system. The robot takes actions that change the environment; the resulting sensory observations then affect future actions. The system can modify its behaviour as it operates and learn from experience.

The last point was added to include learning techniques as state-of-the-art robotics involves adaptive and data-driven controllers that allow the agent to optimise performance while interacting with the environment, such as reinforcement learning and uncertainty-aware model predictive control (Al-Hamadani et al., 2024). Classical control does not learn from its experience and assumes prior knowledge

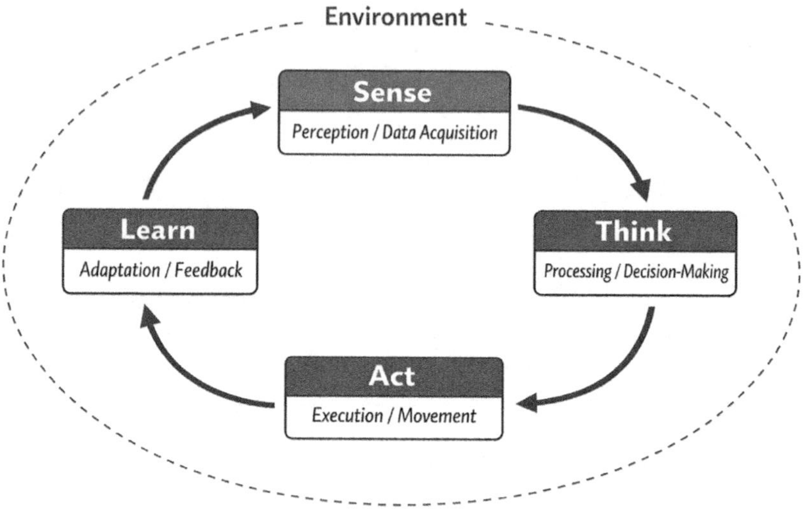

FIGURE 1.1 Sense–Think–Act–Learn model of embodied intelligence.

of the environment; however, intelligent controllers can be updated online to change the behaviour of robots through model predictive control with online model updates or policy adaptation and human feedback (Zhang et al., 2022). Examples of the improvements learning can provide include robustness, adaptability, and task success in real-world applications such as bipedal locomotion, controlling soft robots, and assistive interaction (Radosavovic et al., 2023).

STAL also gives this book a basic architecture that will be used to reference models discussed throughout. Recent publications have adopted similar robotic architectures that separate robot intelligence into sensing, thinking, acting, and learning blocks (Ren & Kim, 2025). Using STAL allows for the easy breakdown of robotics systems to see how each section contributes to intelligent behaviour.

1.4 CASE STUDY: ATLAS ROBOT BY BOSTON DYNAMICS

Boston Dynamics' Atlas humanoid robot system stands out as a prime example of embodied intelligence realised through its hardware and software design for agile whole-body motions (Mikołajczyk et al., 2022). Capable of traversing through rough and unstructured terrain while dynamically carrying out impressive behaviours such as jumping or backflipping, Atlas achieves such agility through an end-to-end sensing–decision–action pipeline (Mikołajczyk et al., 2022; Radosavovic et al., 2023).

Examples of Atlas' sensing include stereo vision cameras, Light Detection and Ranging (LiDAR) sensors, and IMU to sense its surroundings, such as terrain information, obstacles, as well as the robot's orientation with respect to gravity. Atlas can continuously react and adjust its body during locomotion to changes in terrain and disturbances using such sensing capabilities (Miki et al., 2022; Mikołajczyk et al., 2022).

For decision-making, Atlas utilises real-time optimisation and feedback control techniques to stabilise itself and make reactive plans and actions in uncertain environments. Motion planning algorithms and model predictive control are used to rapidly adapt planned motions according to applied forces and contact conditions (De Lima et al., 2024; Dudek et al., 2022; Radosavovic et al., 2023).

Actuation is done through powerful hydraulic actuators that facilitate high-speed and torque-rich motion at various joints across the robot for both locomotion and manipulation capabilities. Reactive control policies are utilised for dynamic balance maintenance and executing agile behaviours such as backflips (Miki et al., 2022; Mikołajczyk et al., 2022; Radosavovic et al., 2023).

Atlas also demonstrates learning abilities as experience interacting with the world can be leveraged to improve performance through techniques such as policy updates or perception–control algorithms (De Lima et al., 2024; Radosavovic et al., 2023). In summary, Atlas exemplifies the STAL schema and currently serves as a leader and gold-standard benchmarking suite for legged robotics.

1.5 COMPONENTS OF INTELLIGENT ROBOTIC SYSTEMS

Architecture varies, but intelligent robotic systems often consist of perception, decision-making, control, and actuation subsystems (Gan et al., 2024; Liu et al., 2025). At the most basic level, these components utilise sensors, software, and computational resources that allow robots to perceive their environment, make decisions, control their actions, and interact with their surroundings (Ergenc, 2022). Machine learning methods are commonly used across these stages.

Perception systems sense and process information about the robot's environment. Algorithms can include object detection, segmentation, prediction, and simultaneous localisation and mapping (SLAM). Sensing modalities can include cameras, LiDAR, tactile sensors, among others. Perception systems allow robots to build internal models or maps of their environment. Recently, perception systems often use deep learning to map raw sensor data (RGB images, point clouds) to useful outputs (semantic maps, traversability maps) for decision-making and planning (Alazeb et al., 2024; Li et al., 2025; Shi et al., 2023; Wang et al., 2024).

Decision-making systems reason about what action should be taken. Methods can range from logic-based systems to motion planning, optimisation, and reinforcement learning. Decision-making enables robots to function in uncertain, changing environments (Gan et al., 2024; Hu et al., 2025).

Control systems translate decisions into actuator commands using control laws. Controllers are typically designed to ensure stable and accurate tracking of commands in the presence of disturbances and uncertainty. Examples include PID controllers, adaptive controllers, and neural controllers (Kondratenko et al., 2022).

Actuation systems allow robots to impact the world. Actuation includes motors, robot arms, grippers, soft actuators, etc. Performance depends on both mechanical design and the control system.

Communication allows robots to interface with humans, other robots, or the cloud. This allows for remote control, supervision, and distributed computing.

Robotic systems will utilise these subsystems in varying configurations and complexity, typically operating in a loop known as the perception–action cycle (Gan et al., 2024; Li et al., 2025). This allows them to follow the STAL paradigm.

1.6 PRACTICAL INSIGHT: UNDERSTANDING A SIMPLE ROBOT SYSTEM

Autonomous robots are a fitting example to better illustrate embodied intelligence. A simple learning robot such as the line-following robot clearly depicts sensing, control logic, and actuation (Brigido & de Oliveira, 2025; Mane, 2024). Educational line-following robots are cheap and easy to build, which makes them popular platforms for learning and prototyping tasks, while still demonstrating core aspects of autonomous systems (Mane, 2024; Rakesh, 2025). A variety of sensing and actuation systems can be used, but even basic implementations require both to function autonomously (Brigido & de Oliveira, 2025; Ren & Kim, 2025).

During sensing, most robots will employ infrared (IR) proximity sensors to sense a contrast between the line and the ground. Models may also have ultrasonic sensors to avoid obstacles (Mane, 2024; Somesh & Aruna, 2025). This sensor data, although raw to users such as reflectance values, provides information such as where the robot is relative to the line. Derived sensor information can then be used to roughly estimate an offset or deviation from the line to follow (Kondratenko et al., 2022; Rakesh, 2025).

Thinking is taking in that sensed information and translating it into an actionable output. The simplest implementations will utilise threshold-based or rule-based algorithms to determine if the robot should move left or right to return to the line. Newer or advanced methods can utilise proportional control or model-based processes to calculate control signals (Cao et al., 2024; Mane, 2024; Wang et al., 2023; Wang et al., 2024). Essentially, this stage interprets perceived error and creates a corresponding action.

Acting can be defined as the physical output of the robot, in this case, movement along the line. Motors supply movement to each of the wheel assemblies. Motor drivers then take in pulse-width modulation signals to vary speed. Adjustments can be made continuously to keep the robot upright while following curves or rough terrain (Cao et al., 2024; Mane, 2024; Rakesh, 2025; Somesh & Aruna, 2025; Wang et al., 2023).

Learning can be implemented in some systems as well. By observing previous runs, the robot can tune control parameters to increase smoothness of operation over time. This completes the STAL cycle for robotics applications (Cao et al., 2024; Ren & Kim, 2025; Rybczak et al., 2024; Wang et al., 2023).

Overall, even basic robots show how sensing, thinking, and acting work together cyclically to allow intelligence to manifest through the body of the robot (Rybczak et al., 2024).

1.7 CHALLENGES IN EMBODIED INTELLIGENCE

Physical systems are often more challenging environments for control than closed-loop digital environments. Physical systems often have complex geometry, uncertainty, and partial observability (Liu et al., 2024; Mon-Williams et al., 2025). They may also be noisy, dynamic, and nonstationary. Distribution shifts and partial information can cause robot failures, so safety and reliability are important considerations (Lee et al., 2023).

Sensor noise can also degrade system perception. Occlusions, changing lighting conditions, and clutter introduce noise to sensors, forcing robots to make decisions under uncertainty (Lee et al., 2023). Robust perception strategies may include probabilistic estimation, sensor fusion, and filtering (Yang et al., 2024).

Delayed feedback is another consideration. Sensing, computation, or communication delays may cause incorrect actions or damage to the robot or its environment. Solutions for real-time control include edge computing, digital twins, and model predictive control to reduce delays (Mtowe et al., 2025).

Limited actuation can also limit performance. Battery capacity, actuators, and structural components limit robots' speed, force, accuracy, and runtime. Understanding and improving these trade-offs motivates the co-design of robot mechanics, sensing, and control (Holzinger et al., 2022; Liu et al., 2025; Mengaldo et al., 2022).

Deployment also considers environmental variation. Robots may experience changes in parameters not accounted for during training. Domain randomisation, sim-to-real transfer, and continual learning can help address these distribution shifts (Liu et al., 2024; Liu et al., 2025).

Finally, robotics problems are often complex systems that must integrate perception, control, and decision-making subsystems. Each system requires efficient computation and coordination between modules.

1.8 ETHICAL AND REAL-WORLD CONSIDERATIONS

Ethical and social considerations are paramount when designing and deploying embodied intelligent systems. Since these systems directly interact with the physical world and impact human lives, several ethical challenges arise (Huang et al., 2023). Safety, trustworthiness, accountability, and human–robot interaction (HRI) are critical considerations that demand responsible AI and human-centred design practices (Abbo et al., 2025; Camilleri, 2023).

Safety concerns become significant when robots operate in environments frequented by humans or animals. It is imperative to design robots that minimise potential risks, adhere to safety standards, and incorporate fail-safe mechanisms. Robots should respond predictably and safely to unforeseen circumstances to avoid harm or injuries (Giallanza et al., 2024; Panagou et al., 2023; Vecellio Segate & Daly, 2023). Safety is a major concern in various applications, including healthcare and autonomous vehicles (Tóth et al., 2022).

Trustworthy and reliable robot behaviour is crucial to ensure user acceptance. Robots should perform actions predictably and consistently; otherwise, humans tend not to trust their behaviour (Panagou et al., 2023). Transparent communication of robot actions and incorporating explainable AI can enhance trust, particularly in safety-critical systems (Adebayo et al., 2024; Chukwunweike et al., 2024; LeTendre & Gray, 2023).

Accountability and responsibility questions arise as robots and AI systems gain autonomy in decision-making. Assigning blame when something goes wrong with an autonomous system can be challenging. Ethical considerations demand clear governance frameworks that ensure transparency and traceability in AI decision-making processes (Camilleri, 2023; Tóth et al., 2022).

HRI should prioritise usability, predictability, and ethical considerations. Robots should align with human values, including respecting privacy, autonomy, and human dignity (Abbo et al., 2025; Panagou et al., 2023). Social and emotional aspects also play a role in human acceptance of robots, particularly in care-related scenarios (Cantucci et al., 2023; LeTendre & Gray, 2023).

1.9 CONCLUSION

In this chapter, students learned about the concept of embodied intelligence as a starting point for approaching AI from the ground up. They were introduced to how components such as sensors, control systems, and actuators work together holistically to create intelligent systems which behave intelligently through interaction with the environment around them. The STAL framework was also introduced as a means of breaking down the different aspects involved in modelling intelligent behaviour. Students were shown examples of simple robotic systems, as well as more complex systems such as the Atlas robot by Boston Dynamics, that demonstrate these concepts. Students should now have a good grasp of embodied intelligence and how it works. Students should also be able to understand some of the major challenges and concerns faced when designing these types of systems. Intelligent system design will be explored further in the following chapters, with a focus on mathematical models, programming, and control.

1.10 DISCUSSION QUESTIONS

1. What do we mean when we say intelligence is embodied?
2. Describe the function of sensors, actuators, and control systems in robots.
3. Explain how the Sense–Think–Act–Learn framework can be used to analyse a robotic system.
4. Provide an example of embodied intelligence in a real-world environment.
5. What are some major challenges we face when designing embodied intelligent systems?

1.11 EXERCISES

Exercise 1: Finding Examples of Sensors, Actuators, and Controllers in Real Systems

Find an example of a real robot (such as a vacuum robot or a delivery robot). What sensors does it use? What actuators does it use? What is its control system? How do these things work together?

Exercise 2: Mapping out STAL for a Robot

Pick an example of a robot performing a simple task and map out each step of the Sense–Think–Act–Learn framework. For example, how does the robot know to perform the task? How does it know what actions to take? How does it learn from its experiences?

Exercise 3: Writing Out Basic Robot Logic

Write out the logic that a robot would use to follow a line. What if it starts to move away from the line? What should it do?

Exercise 4: Embodied vs. Our Grandfathers' Robots

How is embodied intelligence different from intelligence that we see purely in software (such as your phone)? What are some pros and cons of each?

REFERENCES

Abbo, G. A., Belpaeme, T., & Spitale, M. (2025). Concerns and values in human-robot interactions: A focus on social robotics. *International Journal of Social Robotics*. Advance online publication.

Adebayo, A. S., Ajayi, O., & Chukwurah, N. (2024). Explainable AI in robotics: A critical review and implementation strategies for transparent decision-making. *Journal of Frontiers in Multidisciplinary Research, 3*(1), 45–67.

Alazeb, A., Chughtai, B. R., Al Mudawi, N., Alqahtani, Y., Alonazi, M., Aljuaid, H., Jalal, A., & Liu, H. (2024). Remote intelligent perception system for multi-object detection. *Frontiers in Neurorobotics, 18*, 1386323.

Al-Hamadani, M. N. A., Fadhel, M. A., Alzubaidi, L., & Harangi, B. (2024). Reinforcement learning algorithms and applications in healthcare and robotics: A comprehensive and systematic review. *Sensors, 24*(8), 2461.

Brigido, W. J. H., & de Oliveira, J. M. P. (2025). The line follower robot: A meta-analytic approach. *PeerJ Computer Science, 11*, e1802.

Camilleri, M. A. (2023). Artificial intelligence governance: Ethical considerations and implications for social responsibility. *Expert Systems, 40*(7), e13344.

Cantucci, F., Falcone, R., & Marini, M. (2023). Redefining user expectations: The impact of adjustable social autonomy in human–robot interaction. *Electronics, 12*(1), 215.

Cao, Y., Ni, K., Kawaguchi, T., & Hashimoto, S. (2024). Path following for autonomous mobile robots with deep reinforcement learning. *Sensors, 24*(1), 1–21.

Chukwunweike, J., Lawal, O. A., Arogundade, J. B., & Alade, B. (2024). Navigating ethical challenges of explainable AI in autonomous systems. *International Journal of Science and Research Archive, 9*(2), 101–118.

De Lima, C. R., Khan, S. G., Tufail, M., Shah, S. H., & de Albuquerque Máximo, M. R. O. (2024). Humanoid robot motion planning approaches: A survey. *Journal of Intelligent & Robotic Systems, 110*(2), 35.

Dudek, P., Richardson, T., Bose, L., Carey, S., Chen, J., Greatwood, C., Liu, Y., & Mayol-Cuevas, W. (2022). Sensor-level computer vision with pixel processor arrays for agile robots. *Science Robotics, 7*(67), eabm4052.

Ergenc, C. (2022). A review of art and real-world applications of intelligent perception systems. *Advances in Intelligent Systems and Technologies*, *6*(1), 45–66.

Gan, Y., Zhang, B., Shao, J., Han, Z., Li, A., & Dai, X. (2024). Embodied intelligence: Bionic robot controller integrating environment perception, autonomous planning, and motion control. *IEEE Robotics and Automation Letters*, *9*(3), 2590–2597.

Giallanza, A., La Scalia, G., Micale, R., & La Fata, C. M. (2024). Occupational health and safety issues in human-robot collaboration: State of the art and open challenges. *Safety Science*, *167*, 106313.

Holzinger, A., Saranti, A., Angerschmid, A., Retzlaff, C., Gronauer, A., Pejaković, V., Medel-Jiménez, F., Krexner, T., Gollob, C., & Stampfer, K. (2022). Digital transformation in smart farm and forest operations needs human-centered AI: Challenges and future directions. *Sensors*, *22*(7), 2596.

Hu, J., Wang, Y., Cheng, S., Xu, J., Wang, N., Fu, B., Ning, Z., Li, J., Chen, H., Feng, C., & Zhang, Y. (2025). A survey of decision-making and planning methods for self-driving vehicles. *Frontiers in Neurorobotics*, *19*, 1387652.

Huang, C., Zhang, Z., Mao, B., & Yao, X. (2023). An overview of artificial intelligence ethics. *IEEE Transactions on Artificial Intelligence*, *4*(4), 701–720.

Kondratenko, Y., Atamanyuk, I., Sidenko, I., Kondratenko, G., & Sichevskyi, S. (2022). Machine learning techniques for increasing efficiency of the robot's sensor and control information processing. *Sensors*, *22*(3), 1174.

Kortman, V. G., Mazzolai, B., Sakes, A., & Jovanova, J. (2024). Perspectives on intelligence in soft robotics. *Advanced Intelligent Systems*, *6*(9), 2400091.

Lee, D., Jung, M., Yang, W., & Kim, A. (2023). LiDAR odometry survey: Recent advancements and remaining challenges. *Intelligent Service Robotics*, *16*(4), 711–734.

LeTendre, G., & Gray, R. (2023). Social robots in a project-based learning environment: Adolescent understanding of robot–human interactions. *Journal of Computer Assisted Learning*, *39*(5), 1245–1260.

Li, J., Xu, Z., Li, N., Zhang, K., Xiong, G., Sun, M., Hou, C., Ji, J., Zhang, F., Zhong, J., & Huang, Y. (2025). AI-embodied multi-modal flexible electronic robots with programmable sensing, actuating and self-learning. *Nature Communications*, *16*, 5672.

Liu, R., Hu, X., Jiang, Z., Wang, J., & Zhang, W. (2025). Survey on heterogeneous aquatic robot systems: Communication, perception, navigation, control, decision-making and energy management. *Robot Learning*, *4*(1), 1–27.

Liu, Y., Chen, W., Bai, Y., Luo, J.-H., Song, X., Jiang, K., Li, Z., Zhao, G., Lin, J., Li, G., Gao, W., & Lin, L. (2024). Aligning cyber space with physical world: A comprehensive survey on embodied AI. *IEEE/ASME Transactions on Mechatronics*, *29*(5), 3451–3474.

Mane, I. (2024). Multi-purpose tracker robot. *International Journal for Research in Applied Science and Engineering Technology*, *12*(4), 1486–1492.

Mengaldo, G., Renda, F., Brunton, S. L., Bächer, M., Calisti, M., Duriez, C., Chirikjian, G. S., & Laschi, C. (2022). A concise guide to modelling the physics of embodied intelligence in soft robotics. *Nature Reviews Physics*, *4*(8), 471–489.

Miki, T., Hwangbo, J., Wellhausen, L., Koltun, V., & Hutter, M. (2022). Learning robust perceptive locomotion for quadrupedal robots in the wild. *Science Robotics*, *7*(62), eabk2822.

Mikołajczyk, T., Mikołajewska, E., Al-Shuka, H., Malinowski, T., Kłodowski, A., Pimenov, D. Y., Pączkowski, T., Hu, F., Giasin, K., Mikołajewski, D., & Macko, M. (2022). Recent advances in bipedal walking robots: Review of gait, drive, sensors and control systems. *Sensors*, *22*(11), 4186.

Mon-Williams, R., Li, G., Long, R., Du, W., & Lucas, C. (2025). Embodied large language models enable robots to complete complex tasks in unpredictable environments. *Nature Machine Intelligence*, *7*(3), 245–258.

Mtowe, D. P., Long, L., & Kim, D. M. (2025). Low-latency edge-enabled digital twin system for multi-robot collision avoidance and remote control. *Sensors*, *25*(8), 3105.

Panagou, S., Neumann, W. P., & Fruggiero, F. (2023). A scoping review of human robot interaction research towards Industry 5.0 human-centric workplaces. *International Journal of Production Research*, *61*(12), 4048–4070.

Radosavovic, I., Xiao, T., Zhang, B., Darrell, T., Malik, J., & Sreenath, K. (2023). Real-world humanoid locomotion with reinforcement learning. *Science Robotics*, *8*(75), eadi9343.

Rakesh, M. D. (2025). Automated line-following robot with remote speed control using Blynk. *International Journal for Research in Applied Science and Engineering Technology*, *13*(5), 2325–2333.

Ren, Z., & Kim, J. I. (2025). The role of AI in on-site construction robotics: A state-of-the-art review using the sense–think–act framework. *Buildings*, *15*(7), 1456.

Rybczak, M., Popowniak, N., & Lazarowska, A. (2024). A survey of machine learning approaches for mobile robot control. *Robotics*, *13*(1), 4.

Shi, Q., Sun, Z., Le, X., Xie, J., & Lee, C. (2023). Soft robotic perception system with ultrasonic auto-positioning and multimodal sensory intelligence. *ACS Nano*, *17*(5), 4563–4575.

Somesh, D. H., & Aruna, R. (2025). Line follower robot. *International Journal of Advanced Research in Science, Communication and Technology*, *6*(4), 120–127.

Thaker, R. K. (2024). Integrated deep learning architectures for perception, control, and decision-making in robotics: A framework for sensing, cognition, and transparent decision-making. *International Journal for Multidisciplinary Research*, *6*(4), 55–78.

Tóth, Z., Caruana, R., Gruber, T., & Loebbecke, C. (2022). The dawn of the AI robots: Towards a new framework of AI robot accountability. *Journal of Business Ethics*, *182*(3), 657–679.

Vecellio Segate, R., & Daly, A. (2023). Encoding the enforcement of safety standards into smart robots to harness their computing sophistication and collaborative potential: A legal risk assessment for European Union policymakers. *European Journal of Risk Regulation*, *14*(4), 780–805.

Wang, J., Fader, M. T. H., & Marshall, J. A. (2023). Learning-based model predictive control for improved mobile robot path following using Gaussian processes and feedback linearization. *Journal of Field Robotics*, *40*(1), 3–27.

Wang, T., Jin, T., Lin, W., Lin, Y., Liu, H., Yue, T., Tian, Y., Li, L., Zhang, Q., & Lee, C. (2024). Multimodal sensors enabled autonomous soft robotic system with self-adaptive manipulation. *ACS Nano*, *18*(4), 5678–5692.

Wang, X., Wei, R., Chen, Z., Pang, H., Li, H., Yang, Y., Hua, Q., & Shen, G. (2025). Bioinspired intelligent soft robotics: From multidisciplinary integration to next-generation intelligence. *Advanced Science*, *12*(18), 2401234.

Yang, H., Ding, S., Wang, J., Sun, S., Swaminathan, R., Ng, S. W. L., Pan, X., & Ho, G. W. (2024). Computational design of ultra-robust strain sensors for soft robot perception and autonomy. *Nature Communications*, *15*, 1690.

Zhang, W., Ott, L., Tognon, M., & Siegwart, R. (2022). Learning variable impedance control for aerial sliding on uneven heterogeneous surfaces by proprioceptive and tactile sensing. *IEEE Robotics and Automation Letters*, *7*(3), 7423–7430.

Zhao, Z., Li, W., Li, Y., Liu, T., Li, B., Wang, M., Du, K., Liu, H., Li, Y., Wang, Q., Althoefer, K., & Zhu, S.-C. (2024). Embedding high-resolution touch across robotic hands enables adaptive human-like grasping. *Nature Machine Intelligence*, *6*(12), 1450–1463.

2 Mathematical Thinking for Intelligent Systems

2.1 INTRODUCTION

Robotics is built on mathematics: the controller design techniques we employ at every level of abstraction depend crucially on mathematics, whether you are doing motion planning or perception algorithms (Mengaldo et al., 2022). Mathematics allows us to model, specify, and verify the kinematics, dynamics, and behaviour of robots in the presence of uncertainty (Lauri et al., 2022). These models give robots certifiable guarantees about safety and performance as they carry out their tasks (Lauri et al., 2022). Robots are electromechanical systems after all. But it is mathematics that makes them move intelligently and reliably. Mathematics allows them to walk, grasp objects, navigate to destinations, and avoid obstacles (Löw & Calinon, 2022).

Mathematics also allows robots to see and reason about the world around them. Optimal control, estimation, and probabilistic decision-making are examples of mathematical techniques that allow robots to reason under uncertainty from noisy sensor information and then take effective actions in the real world (Liu et al., 2023; Wensing et al., 2022). Robots rely on mathematics to describe the physical quantities that we care about, such as position, velocity, orientation, and forces applied to and by the system. Mathematical equations describe how these quantities will evolve over time and how the system should act.

A controller, for example, will mathematically specify how the inputs to the system should change to achieve the desired behaviour (e.g., stable, follow a certain trajectory), in the presence of disturbances that alter the system's behaviour (Liu et al., 2023).

This chapter provides an introduction to some of the mathematics that is used in robotics, such as algebra, and shows how these mathematical tools can be applied to coordinate transformations, plan motions, and make decisions under uncertainty.

2.2 FOUNDATIONS OF MATHEMATICAL THINKING IN ROBOTICS

Mathematical thinking broadly refers to representing, analysing, and predicting systems' behaviour using equations, functions, and models as tools (Lauri et al., 2022; Mengaldo et al., 2022; Yang et al., 2025). While this encompasses numerical abilities, mathematics more generally captures relationships between variables or other systems' states changing through time or input. Whether written out analytically or approximated through computation, mathematics serves as a mechanism for abstractly modelling relationships in robots (Kusaka & Tanaka, 2025; Wensing et al., 2022). This intuition can take forms like those found in control theory: inputs, states, outputs.

DOI: 10.1201/9781003758501-3

Modelling physical systems through mathematics sits at the core of mathematical thinking. Simply put, modelling is establishing mathematical relationships which describe how something works. For robots, this can mean modelling how the robot moves or forces propagate through it, how the robot will react to the environment, or any other aspect of its physical behaviour (Mengaldo et al., 2022). The position of a joint may be written as a function of time, while its velocity is its derivative. Relationships such as these, when written down as differential equations or kinematic chains, allow us to precisely describe motion and infer physical properties of the system like stiffness or inertia from observed sensor data.

Variables and parameters are another concept related to this idea. Variables describe quantities which evolve through time: position, velocity, and battery life. Parameters describe properties of the system either too difficult or slow to measure that affect its behaviour such as motor limits, inertia, or gains in a controller (Kusaka & Tanaka, 2025).

Error, or the difference between some desired and current state, also plays an important role in mathematics. For robots, controllers provide ways of thinking about how to act on a system to reduce error. Many control techniques, such as PID (Proportional–Integral–Derivative) control or model predictive control, allow users to provide stability, accuracy, and safe interaction guarantees (Tian et al., 2024).

Mathematical thinking ultimately allows users to think about how something will behave before taking an action. By analysing a model of a system, we can reason about how it will behave and plan for it. Abstractions like these allow for advanced robot behaviours like locomotion, manipulation, and safe human–robot interaction.

2.3 MODELLING MOTION AND CONTROL IN ROBOTIC SYSTEMS

Motion modelling and control involve characterising how robots move in response to actuation commands, as well as control inputs needed to follow trajectories and resist perturbations in uncertain environments (Meduri et al., 2022). Mathematical representations are used to define how the pose of a robot changes over time; kinematic and dynamic modelling rely on sets of equations parameterised by state variables that typically follow Euler–Lagrange or Newton–Euler formulations to characterise the behaviour of robot systems (Kong et al., 2023). Modelling techniques enable closed-form representations of systems to reason about their behaviour and synthesise controllers that can meet guarantees such as tracking accuracy and robustness to disturbances and modelling errors (Schegg & Duriez, 2022).

In its simplest form, robot motion can be characterised using position, velocity, and acceleration. These variables define the fundamental relationships used to model robot motion using equations of kinematics or dynamics (Kong et al., 2023). Position is a numerical value that specifies the location of the robot; velocity is the rate of displacement over time, and acceleration is the rate of velocity. The quantities are inherently related via integrals and derivatives and can be used to represent motion in linear or rotational coordinates for both manipulators and mobile robots/platforms (Meduri et al., 2022). For instance, motion of a mobile robot following a prescribed trajectory must be aware of its position as it travels; this relationship can be modelled by defining the robot's commanded velocities and linking it to its pose evolution to track the desired path (Miranda-Colorado, 2022).

Error is another important concept commonly used in controls, which mathematically represents the difference between where the system should be and where the system is; many trajectory tracking control designs are developed using this concept (Azeez et al., 2023). For example, consider driving in a straight line but starting to turn away from the desired direction. There is an instantaneous error between where you are and where you should be, and you can make use of this error to steer back to your desired path. This concept can be formalised as a tracking error of your position or orientation and used as the basis for control (Seleem et al., 2023). Mathematical models allow this error to be represented in useful ways to design controllers that can guarantee stability of the system and allow rigorous analysis.

The error can be used by control systems to modify the inputs to the robot to reduce this error, closing the loop between sensing and actuation on the robot (Miranda-Colorado, 2022). PID control is a commonly used control model in robotics and remains a ubiquitous baseline in both industrial and academic settings (Massoud & Libby, 2023). PID controllers use a linear combination of three control techniques to control the system.

$$u(t) = K_p e(t) + K_i \int e(t)\,dt + K_d \frac{de(t)}{dt}$$

The control law describes how the control inputs will react to changes in error. The *e(t)* term is often referred to as error at time t. The proportional term corrects proportional to the current error, the integral term will compensate for how much error has accumulated in the past, and the derivative term will attempt to estimate future error based on its current rate of change; tuning of the gains will affect tracking performance of the system (Azeez et al., 2023). The combination of these elements will allow for high response speeds, low steady-state error, and dampening of overshoots. Many articles have recently been published improving upon PID gains, as well as intelligent control and adaptive control methods to increase PID's robustness (Massoud & Libby, 2023; Yang et al., 2024).

PID controllers describe motion as the relationship between control and motion updates continually; these models and controllers are linked. The robot senses where it is, determines where it wants to go based on that information, and uses a controller to calculate and correct for the error between where it is and where it wants to go (Schegg & Duriez, 2022). This updates frequently allowing robots to deal with unpredictable conditions. Sometimes this control loop runs at rates of 1 kHz or higher. We will expand heavily upon this foundation in future sections, including robot navigation, optimisation, and autonomous systems (Kong et al., 2023). Essentially, modelling the relationship between motion allows for principled approaches to controller synthesis and lets us analyse how the controller will respond to motion (Meduri et al., 2022). Many times, this allows controllers to be designed that can be guaranteed to stabilise the system in the presence of modelling errors or external disturbances (Zhang et al., 2025). Engineers model these relationships mathematically and use the model to design controllers that will allow the system to respond how they want it to.

Robotic systems to desired specifications. This modelling framework allows incorporation of both model-based and learning-enhanced controllers to deal with complex environments and tasks. Understanding how motion works allows us to do something similar for control.

2.4 CASE STUDY: PID CONTROL IN MOBILE ROBOTS

PID control is a popular intuitive robust and computationally inexpensive method to stabilise manipulators, wheeled robots, and underactuated platforms (Benotsmane & Kovács, 2023). PID control serves as a common baseline in robotics research and education and serves as a friendly introduction to feedback control by tying mathematical concepts to physical behaviour (Azeez et al., 2023). PID controllers are frequently taught using line following, speed control, or trajectory tracking applications (Nazir et al., 2024).

Mobile robots must follow a desired heading or trajectory specified as a function of time. Position sensors like wheel encoders, inertial measurement units, or vision sensors are used to provide feedback on the robot's current state estimate (Pussente et al., 2023). The error between desired and current state is fed as the input to the PID controller (Benotsmane & Kovács, 2023).

The proportional controller creates an output value that is proportional to the current error. This allows the controller to respond quickly to positional errors but can leave a constant steady-state error. This error is present when the system cannot reach the setpoint due to disturbances or modelling errors (Riaz et al., 2023).

The integral controller slowly increases controller output based on the cumulative error over time. This makes the integral controller good at dealing with persistent offsets, such as a constant bias or noise source (Azeez et al., 2023). The derivative controller reacts to how fast the error is changing. Derivative control can increase the stability of a system and reduce overshoot by damping the output response.

PID controllers form a stable closed-loop controller when used together. Consider a line following robot. Positional error is used as input to a PID controller to change the speeds of the robot's wheels.

2.5 CORE MATHEMATICAL TOOLS FOR ROBOTICS

Mathematical definitions allow engineers to design intelligent machines that behave reliably in the real world. Robots are first modelled using mathematics to describe their motion and dynamics. Engineers can then analyse these models and translate them into code that runs on a computer (Mengaldo et al., 2022). Mathematics can describe robot dynamics physically and algorithmically. Fundamental concepts in robotics such as position, orientation, and forces all use mathematics. By abstracting real-world problems into mathematical representations, we can solve for systems ranging from industrial manipulators to mobile and soft robots (Liu et al., 2023).

2.5.1 Control Systems Algebra

Control engineers use algebra to describe mathematical equations relating variables to each other. Kinematics equations, dynamics equations, and controller equations are often written using algebraic notation (Kusaka & Tanaka, 2025). Variables can represent system states, error values, and control inputs. Statements using variables can be manipulated to express control laws. The difference between where a robot should be and where it can be written algebraically. PID controllers and optimisation-based controllers use this error to calculate a control input signal (Yaren & Kizir, 2025). Algebra serves as the backbone for both model-based control and data-driven control. Algebraic equations can be symbolically represented and solved or approximated using numerical computation in programming languages (Löw & Calinon, 2022).

2.5.2 Trigonometry Robot Position and Heading

Trigonometry is the study of angles, rotations, and relationships between lengths. Trigonometry is heavily used to represent a robot's heading or how it should move between points. Many systems use trigonometry to describe how to face towards a certain point or direction (Tang et al., 2024). Consider a mobile robot navigating to a point in two-dimensional space. The robot has an angular quantity that it can manipulate and measure. Trigonometric functions allow us to map angles to x and y positions. Maps of these functions, known as sine and cosine, are heavily used in robotics for computing headings and robot motion. This concept is used every time a robot plans a path or localises itself (Lauri et al., 2022).

2.5.3 Linear Algebra Vectors and Matrices

Linear algebra defines many of the mathematical objects we have already mentioned: vectors and matrices. A robot exists in at least two dimensions on a horizontal plane and three when including elevation. Linear algebra allows us to express a robot's position, orientation, and motion within these spaces. Rotation transformations, translation transformations, and scaling transformations can be represented with matrices. Magnitude with direction can be represented using vectors. Concepts from linear algebra are used anytime we represent coordinate transformations, kinematics, or computer vision.

2.6 PRACTICAL INSIGHT: SIMULATING ROBOT MOTION

Simulation allows students or engineers to visualise the behaviour of a mathematical model as if it were interacting with the physical world (Sujatha et al., 2025). Simulations let you validate controllers, prototype designs, and see how a system might respond to different trajectories, loads, or other disturbances (Ding et al., 2022). Robot manufacturers implement equations of motion that adhere to physical laws into simulation environments, which allow you to mimic robot behaviour before purchasing and assembling hardware yourself (Saldarriaga-Mesa et al.,

2025). Overall, simulation mitigates many risks and time costs associated with robotics.

Motion of a robot in simulation is often computed through numerical integration. This process can be thought of as breaking up time into small intervals, or steps. Then, at every step, you can use mathematical equations that describe how the state of your system changes to update variables like position, velocity, or orientation (Strashnov & Mikhaylyuk, 2025). A simple example is position equals the previous position plus velocity times the timestep. You can extend this concept to account for rotation, acceleration, contact forces, friction, and soft robot deformable dynamics (Ding et al., 2022).

Simulation environments are also commonly used to test control systems like PID controllers. Simulation allows you to change controller parameters and see how your robot may react to certain inputs or external disturbances. Maybe you want your robot to respond faster to errors, so you increase proportional gain. But then your simulation crashes because your robot goes unstable. Or maybe you want to reduce how much your robot oscillates around a setpoint, which you know you can do by increasing derivative action. All these questions can be answered with the help of simulation.

Most simulation environments operate on the concept of discrete time using loops. During each iteration of a loop, your simulated robot will take in sensed values, compute errors based on desired values, and update controls to act on (Escandon-Esparza & Jurado, 2025). This is very similar to how actual robots operate.

Simulation allows you to develop intuition about how your system works. By tweaking parameters and observing how the simulation responds, you will begin to connect the dots between the mathematics and real-world behaviour.

2.7 CHALLENGES IN MATHEMATICAL MODELLING

Mathematical models never perfectly represent reality. Physical phenomena can be complex, nonlinear, and uncertain, requiring simplifying assumptions for mathematical tractability. As a result, all models are wrong to some degree (Lauri et al., 2022). Uncertainty in models can come from unknown parameters, unmodelled dynamics, and changing operating conditions. These differences between a model and its real-world counterpart can result in unpredictable or undesired behaviour (Street et al., 2023). Modelling errors are particularly difficult for highly coupled systems like soft robots (Zheng & Lin, 2022).

Computational errors can also accumulate over time. If the control actions at each step are off by a little bit, the errors may compound until the robot takes large and sudden deviations from its target state. This can lead to errors like drift or poor tracking (Gonzalez Riascos et al., 2025). For instance, if your mathematical model of a drone is slightly off when computing velocity or facing, the drone may start to drift away over time. System behaviour is often sensitive to initialisation conditions and parameters as well.

Sensor noise presents similar issues, as noisy data can result in incorrect calculations. Measurements of position, velocity, etc. are never perfect and are typically affected by noise, interference, and environmental conditions like temperature or vibration. This means they are not as accurate or reliable as assumed in mathematical models. Many modelling and control methods assume perfect knowledge of these inputs, so noisy data can cause failure or unexpected behaviour. Fortunately, there are tools like observers and various filtering techniques to help deal with noisy measurements and estimate the true state of the system (Gonzalez Riascos et al., 2025).

Mathematical models also tend to be simplified to make them easier to compute. Friction, delays, external forces, etc. are often ignored to simplify equations. While these decisions can speed up computations for analysis and control, they create a "reality gap" between how the mathematical model should behave and the actual system.

Some control parameters can also be difficult to tune. Improperly tuned controllers may become unstable or respond slowly to commands. For all these reasons, adaptive, robust, and learning-based control methods have become popular in modern robotics.

2.8 ETHICAL AND REAL-WORLD CONSIDERATIONS

Intelligent systems must behave safely, even if the mathematical model they operate under is wrong. Mathematical models are simplifications of reality, and depending on them too much can cause problems in physical settings that don't behave the way they expect (Alzubaidi et al., 2023). For control specifically, and safety critical applications especially, there are some unique considerations.

Misalignments in assumed conditions and reality are one concern. If control is built on top of or tuned for an inaccurate model of the system it should control, or if a learning system is trained with data that does not represent its environment well, then it may perform unexpected or unsafe behaviour when run in the real world. Safety critical applications like autonomous cars and medical usage are particularly important cases here (Alzubaidi et al., 2023).

Error accumulation can happen if small modelling mismatches or computational errors cause incorrect output from the control system, which are then amplified as they propagate through the control process. In software with no human oversight, such as autonomous systems, this can cause unintended behaviour by deciding on an incorrect or inappropriate action. Validating models before deployment, careful monitoring of systems, and putting safety critical systems in a position to fail safely can help address this concern (Joshi, 2025).

Limited interpretability of control systems is also a challenge. Some control systems, particularly ones that use optimisation or machine learning, may operate in ways that are impossible for a human to understand. The resulting lack of explainability can cause issues with trust or accountability of these systems, and research into methods of explainable AI is growing to address these issues. Regulation surrounding these problems is still in early stages as well (Chaudhary, 2024).

Considerations of bias and fairness come up as well. Mathematical models may encode unintended biases from data they learn from or assumptions made during design. This can lead to unfair treatment of certain groups from decisions made by the control system. Applications within robotics that interact with humans and provide services to them can face similar issues if the robots' behaviour learns to favour certain groups over others (Sreerama & Krishnamoorthy, 2022).

Control systems should ideally always be validated to behave correctly and safely even outside of mathematical models and idealised conditions (Alzubaidi et al., 2023).

2.9 CONCLUSION

The chapter introduced mathematics as the common underlying framework and language through which model-based and learning-augmented intelligent robotic systems are conceptualised, designed, programmed and controlled, which aligns with current practice (Wensing et al., 2022). It showcased how mathematical thinking facilitates the description, analysis, and control of system behaviour using models and relationships, ranging from kinematics and dynamics to model-based optimisation controllers (Löw & Calinon, 2022; Mengaldo et al., 2022). Relevant mathematical tools were introduced with particular focus on algebra, trigonometry, linear algebra and calculus to lay the foundations for describing motion, orientation and system dynamics of rigid and soft robots moving in three-dimensional space (Schegg & Duriez, 2022).

Mathematical modelling for control was further emphasised through the example of PID control of mobile robots, since PID control remains a popular baseline method for controller evaluation in experimental and industrial robotics today (Villalba-Aguilera et al., 2025). Students should now appreciate how mathematical constructs such as error, integration, and rate-of-change affect robot behaviour in terms of stability, tracking performance, disturbance rejection, etc., both through simulation and their newfound intuition. This also aligns with current practices of validating model-based controllers (may be learning-based or optimisation-based) through simulation (Yaren & Kizir, 2025). Moreover, students should recognise that mathematical models are only approximations of the real-world, and should strive for models that are accurate, validated, and capable of adapting to uncertainty, unmodelled dynamics, and soft–environment interactions when deploying in real-world environments (Mengaldo et al., 2022). Students should also consider ethical aspects when building and deploying mathematical models, which aligns with recently proposed frameworks for trustworthy and safe intelligent robotic systems and AI-enabled robotics (Alzubaidi et al., 2023).

These concepts will serve as prerequisites for upcoming chapters that introduce programming languages, system integration, more complex controls and planning algorithms to further enable the design of intelligent and autonomous robotic systems through synergies of model-based, learning-based, and optimisation-based approaches (Chai et al., 2022).

2.10 DISCUSSION QUESTIONS

1. Why do we need math to understand and control smart robots?
2. Describe how variables, equations, and models help us predict robotic behaviour.
3. How does PID control use error to correct the system?
4. How can tiny math errors cause giant robot failures?
5. Models are just approximations of reality. What are some limits of math models?

2.11 EXERCISES

Exercise 1: From Sensors to Math

A robot has a sensor that tells it where it is relative to something it cares about. If you know the desired position and the actual position read from the sensor, calculate the error. How would this error be used in a control system?

Exercise 2: PID in Plain Language

Describe what the P, I, and D do in a PID control system. Give a basic example of how each component works to control system behaviour.

Exercise 3: Moving in a Straight Line

A robot is travelling in a straight line at a constant speed. Determine where the robot would be after a certain amount of time. How would your answer change if the speed was different?

Exercise 4: What's Wrong with This Model?

Models are simplifications of reality. What are some problems with using mathematical models to represent the real world? Give an example.

REFERENCES

Alzubaidi, L., Al-Sabaawi, A., Bai, J., Dukhan, A., Alkenani, A. H., Al-Asadi, A., Alwzwazy, H. A., Manoufali, M., Fadhel, M., Albahri, A. S., Moreira, C., Ouyang, C., Zhang, J., Santamaría, J. I., Salhi, A., Hollman, F., Gupta, A., Duan, Y., Rabczuk, T., Abbosh, A., & Gu, Y. (2023). Towards risk-free trustworthy artificial intelligence: Significance and requirements. *International Journal of Intelligent Systems*, *38*(10), 11876–11926.

Azeez, M. I., Abdelhaleem, A., Elnaggar, S., Moustafa, K., & Atia, K. R. (2023). Optimization of PID trajectory tracking controller for a 3-DOF robotic manipulator using enhanced Artificial Bee Colony algorithm. *Scientific Reports*, *13*, 11207.

Benotsmane, R., & Kovács, G. (2023). Optimization of energy consumption of industrial robots using classical PID and MPC controllers. *Energies*, *16*(8), 3485.

Çakıroğlu, Ü., & Yıldırım, M. (2024). Exploring students' mathematical thinking skills in educational robotics activities. *Thinking Skills and Creativity*, *52*, 101336.

Chai, R., Niu, H., Carrasco, J., Arvin, F., Yin, H., & Lennox, B. (2022). Design and experimental validation of deep reinforcement learning-based fast trajectory planning and control for mobile robot in unknown environment. *IEEE Transactions on Neural Networks and Learning Systems*, *33*(12), 7994–8007.

Chaudhary, G. (2024). Unveiling the black box: Bringing algorithmic transparency to AI. *Masaryk University Journal of Law and Technology*, *18*(1), 65–92.

Ding, L., Niu, L., Su, Y., Yang, H., Liu, G., Gao, H., & Deng, Z. (2022). Dynamic finite element modeling and simulation of soft robots. *Chinese Journal of Mechanical Engineering*, *35*, 33.

Escandon-Esparza, G., & Jurado, F. (2025). Virtual model development and control for an EV3 BallBot robotic system. *Processes*, *13*(8), 1295.

Gonzalez Riascos, J. E., Castaño Vanegas, S. M., Medeiros, V. S., Becker, M., & Castillo García, J. F. (2025). Noise parameter tuning for an invariant extended Kalman filter implementation using genetic algorithms. In *2025 Brazilian Conference on Robotics (CROS)* (pp. 1–6). IEEE.

Joshi, S. (2025). Review of Gen AI models for financial risk management. *International Journal of Scientific Research in Computer Science, Engineering and Information Technology*, *11*(1), 140–150.

Kong, L., He, W., Chen, W., Zhang, H., & Wang, Y. (2023). Dynamic movement primitives based robot skills learning. *Machine Intelligence Research*, *20*(1), 60–76.

Kusaka, T., & Tanaka, T. (2025). A chain rule-based generalized framework for efficient dynamic analysis of complex robotic systems. *Robotics*, *14*(4), 89.

Lauri, M., Hsu, D., & Pajarinen, J. (2022). Partially observable Markov decision processes in robotics: A survey. *IEEE Transactions on Robotics*, *38*(6), 3443–3462.

Liu, J.-Y., Borja, P., & Santina, C. D. (2023). Physics-informed neural networks to model and control robots: A theoretical and experimental investigation. *Advanced Intelligent Systems*, *5*(8), 2300034.

Löw, T., & Calinon, S. (2022). Geometric algebra for optimal control with applications in manipulation tasks. *IEEE Transactions on Robotics*, *38*(6), 3905–3924.

Massoud, M. M., & Libby, J. (2023). Comparative analysis of evolutionary algorithms for PID controller optimization in pneumatic soft robotic systems: A simulation and experimental study. *IEEE Access*, *11*, 128456–128472.

Meduri, A., Shah, P., Viereck, J., Khadiv, M., Havoutis, I., & Righetti, L. (2022). BiConMP: A nonlinear model predictive control framework for whole body motion planning. *IEEE Transactions on Robotics*, *38*(4), 2341–2359.

Mengaldo, G., Renda, F., Brunton, S. L., Bächer, M., Calisti, M., Duriez, C., Chirikjian, G. S., & Laschi, C. (2022). A concise guide to modelling the physics of embodied intelligence in soft robotics. *Nature Reviews Physics*, *4*(9), 595–613.

Miranda-Colorado, R. (2022). Observer-based proportional integral derivative control for trajectory tracking of wheeled mobile robots with kinematic disturbances. *Applied Mathematics and Computation*, *429*, 127254.

Nazir, K., Kim, Y.-W., & Byun, Y. (2024). Predictive PID control for automated guided vehicles using genetic algorithm and machine learning. *IEEE Access*, *12*, 76543–76560.

Pussente, G. A. N., Aguiar, E., Marcato, A., & Pinto, M. (2023). UAV power line tracking control based on a type-2 fuzzy-PID approach. *Robotics*, *12*(2), 41.

Riaz, S., Qi, R., Tutsoy, O., & Iqbal, J. (2023). A novel adaptive PD-type iterative learning control of the PMSM servo system with friction uncertainty in low speeds. *PLoS One*, *18*(1), e0280112.

Saldarriaga-Mesa, B., Varela Aldás, J., Roberti, F., & Toibero, J. (2025). A novel methodology for designing digital models for mobile robots based on model-following simulation in virtual environments. *Robotics*, *14*(9), 124.

Schegg, P., & Duriez, C. (2022). Review on generic methods for mechanical modeling, simulation and control of soft robots. *PLoS One*, *17*(1), e0261635.

Seleem, I. A., El-Hussieny, H., & Ishii, H. (2023). Imitation-based motion planning and control of a multi-section continuum robot interacting with the environment. *IEEE Robotics and Automation Letters*, *8*(3), 1660–1667.

Sreerama, J., & Krishnamoorthy, G. (2022). Ethical considerations in AI: Addressing bias and fairness in machine learning models. *Journal of Knowledge Learning and Science Technology*, *1*(1), 132–142.

Strashnov, E., & Mikhaylyuk, M. V. (2025). Motion simulation in virtual environment of flying robots inside orbital space stations. *Mekhatronika, Avtomatizatsiya, Upravlenie*, *26*(5), 312–321.

Street, C., Mansouri, M., & Lacerda, B. (2023). Formal modelling for multi-robot systems under uncertainty. *Current Robotics Reports*, *4*, 231–244.

Sujatha, A. A., Kolahdooz, A., Jafari, M., & Hajfathalian, A. (2025). Simulation and control of the KUKA KR6 900EX robot in Unity 3D: Advancing industrial automation through virtual environments. *EAI Endorsed Transactions on AI and Robotics*, *9*(1), e5.

Tang, J., Wu, S., Lan, B., Dong, Y., Jin, Y., Tian, G., Zhang, W., & Shi, L. (2024). GMPC: Geometric model predictive control for wheeled mobile robot trajectory tracking. *IEEE Robotics and Automation Letters*, *9*(4), 3555–3562.

Tian, G., Tan, J., Li, B., & Duan, G. (2024). Optimal fully actuated system approach-based trajectory tracking control for robot manipulators. *IEEE Transactions on Cybernetics*. *54*(12):7469–7478. Advance online publication.

Villalba-Aguilera, E., Blesa, J., & Ponsa, P. (2025). Model-based predictive control for position and orientation tracking in a multilayer architecture for a three-wheeled omnidirectional mobile robot. *Robotics*, *14*(3), 72.

Wensing, P. M., Posa, M., Hu, Y., Escande, A., Mansard, N., & Prete, A. (2022). Optimization-based control for dynamic legged robots. *IEEE Transactions on Robotics*, *38*(6), 3339–3359.

Yang, J., Wu, Q., Li, S., Ye, Y., & Luo, C. (2025). Integrated modeling and control optimization of biped wheel-legged robot. *IEEE Robotics and Automation Letters*, *10*(2), 1456–1463.

Yang, Q., Zhang, F., & Wang, C. (2024). Deterministic learning-based neural PID control for nonlinear robotic systems. *IEEE/CAA Journal of Automatica Sinica*, *11*(5), 1130–1144.

Yaren, T., & Kizir, S. (2025). Real-time nonlinear model predictive control of a robotic arm using spatial operator algebra theory. *Journal of Field Robotics*, *42*(1), 3–24.

Zhang, T., Wang, H., Lv, P., Pan, X., Wang, D., Yuan, B., & Yu, H. (2025). Real-time motion generation for robot manipulators in complex dynamic environments. *Advanced Intelligent Systems*, *7*(7), 2400456.

Zheng, T., & Lin, H. (2022). PDE-based dynamic control and estimation of soft robotic arms. In *2022 IEEE 61st Conference on Decision and Control (CDC)* (pp. 4303–4308). IEEE.

3 Patterns, Logic, and Thinking Like a Computer

3.1 INTRODUCTION

Computational thinking means presenting problems and solutions so that a computer or robot will be able to follow the instructions (Barradas et al., 2024). Problem-solving by following the logical steps requires a machine to program computers and design algorithms that can carry out tasks (Sarker, 2022). When solving problems with computational thinking, people think like computers use, which allows for abstraction (Barradas et al., 2024; Sarker, 2022). Computational thinking in robots and AI involves translating thoughts into specific directions/rules.

For computers to perform tasks, instructions need to be provided in a manner that the machine cannot misunderstand. Instructions to artificial intelligence and robots need to be unambiguous because, unlike humans, they cannot use intuition to fill in gaps (Ghzouli et al., 2022; Sarker, 2022). Instructions are provided using programming languages (also unambiguous) and are created using algorithms and constructs such as "if-then" statements. Robots are completely dependent on the instructions they are provided. Robots sense their environment, make decisions, and act by using state machines, behaviour trees, or rule-based systems (Ghzouli et al., 2022). These machines typically use "if-then" rules. "If" something is detected, "then" they will perform a certain action.

Computational instructions break large goals into pieces that a robot can represent with the help of computers. These include loops, if-else statements, variables, and lists to help the robot perform the same action every time (Sarker, 2022). If a robot is cleaning the house and comes across an object in its way, it will need to have some sort of routine. Maybe it will stop and notify you that it requires assistance, maybe it will try to go around the object. But in any case, a rule, a set of rules, or a decision tree will be used to help the robot decide what to do.

By learning about computational thinking, students can understand how problems can be converted into step-by-step instructions for a machine to follow. It will allow them to think about solutions that could allow for AI and computer use to solve problems (Barradas et al., 2024). Students will also become more knowledgeable about how to solve problem in a world filled with AI and automation technology (Amri et al., 2022).

3.2 FOUNDATIONS OF COMPUTATIONAL THINKING

Computational thinking involves solving problems through processes that a computer can understand and execute (Pirzado et al., 2025; Wu et al., 2024). This means that problems must be analysed through a logical structure and simplified to create

DOI: 10.1201/9781003758501-4

smaller representations of the solution that are translatable into computational steps. Computational thinking includes four major principles: decomposition, pattern recognition, abstraction, and algorithms (Ertas Kalay & Ipek, 2024). Robots leverage these principles to solve problems effectively and at scale (Jawawi et al., 2022).

Decomposition focuses on dividing problems into smaller subproblems (Pirzado et al., 2025; Wu et al., 2024). Robots decompose problems into sensing, planning, and control components to design individual pieces that can be implemented and then assembled as a system (Chichekian et al., 2023; Jawawi et al., 2022).

Pattern recognition deals with identifying common elements across problems or environments (Pirzado et al., 2025; Wu et al., 2024). Robots may leverage patterns in the environment to reuse solutions, like reusing the same obstacle-avoidance behaviour in two nearby environments with similar layouts (Amri et al., 2022).

Abstraction is achieved by only caring about certain properties of a problem or environment while ignoring others (Jawawi et al., 2022; Pirzado et al., 2025; Wu et al., 2024). Robots use abstractions to represent objects in the environment in simpler ways, like representing an obstacle as a bounding box with properties like position and size.

Computational thinking through algorithms defines the step-by-step procedures for solving a task (Pirzado et al., 2025; Wu et al., 2024). Algorithms specify the input, output, and computations required to solve a task. Robot behaviours are specified by algorithms, like how to navigate to a goal, avoid obstacles, or move joints (Chichekian et al., 2023).

Through computational thinking, roboticists can design systems that solve problems reliably at scale through well-defined structure and organisation (Ertas Kalay & Ipek, 2024; Wu et al., 2024).

3.3 MODELLING DECISION-MAKING IN INTELLIGENT SYSTEMS

Decision-making processes in intelligent systems can be abstracted and represented using explicit models that take some inputs and specify outputs or actions. Decision trees and tree-based control architectures are commonly applied to represent decision-making processes (Huang, 2024; Sirichanya & Kesorn, 2023). An effective technique is to use a decision tree structure that encodes the system's decision-making process and affords transparent and interpretable reasoning about behaviour (Yuan et al., 2022). Decision trees provide robust models that are popular for robot behaviour generation, where control decisions must be well-defined and repeatable across similar contexts (Bernardo et al., 2023).

Decision trees consist of root, internal, and leaf nodes arranged in a hierarchy that formulate the decision-making process as a sequence of if–then conditions (Huang, 2024; Yuan et al., 2022). Starting from the root, each decision checks some condition based on an input or state (e.g., sensor data) and branches based on a feature threshold or passes a logical test (Sirichanya & Kesorn, 2023). Then, depending on whether the statement is true or false, the corresponding branch is executed towards some action leaf, or the next internal decision node is evaluated (Huang, 2024). This process can be repeated recursively or iteratively until some terminus or action is reached and mimics conditional if–then queries against the input data (Yuan et al.,

2022). Decision trees break complex decisions and behaviours down into interpretable chunks that can still be systematised and optimised.

As an illustrative example of decision-making, a robot could choose to either move forward or change direction based on some sensor input indicating whether an obstacle is present (Bernardo et al., 2023). Specifically, an obstacle detection sensor, such as a proximity sensor or camera, might detect if there is an obstacle in front of the robot (Yu et al., 2024). In this case, a "turn" action may be taken that branches to turning the robot right or left, depending on some further criterion or sensor bias (Bernardo et al., 2023). If no obstacle is detected, some "move forward" action may be taken that keeps the robot travelling straight. Each of these decisions corresponds to a defined condition that is either true or false to guarantee predictable behaviour that can be validated and verified for safety requirements (Yu et al., 2024; Yuan et al., 2022).

Figure 3.1 provides a general schema for binary decision trees, showing how they are divided into nodes that represent successive decision checks (Huang, 2024; Yuan et al., 2022). The root node represents the start state or input that begins the decision-making process, then internal nodes denote decision checks that funnel the decision process towards certain leaves. By reductively checking conditions and eliminating branches, the tree structure permits decision-making to converge on actions (Sirichanya & Kesorn, 2023). Since modifications to decision-making can be made by mutating local nodes, decision trees also allow modular updates to subsets of decisions.

Decision trees are inherently related to flow-control features of programming languages such as conditional statements (if/else) and loops and may be converted to or from such abstract syntax in many cases (Huang, 2024; Yuan et al., 2022). For programmers, decision trees may represent intermediary steps between logical reasoning about decisions and the final implementation in code. Similarly, decision trees may be compiled down into sets of if else statements or more abstract behaviour-tree controllers.

Decision trees have been widely applied to allow robots to respond consistently to inputs or sensor data. Tree-based controllers and variants are often used in autonomous robots, unmanned underwater vehicles, and intelligent service platforms (Bernardo et al., 2023; Yu et al., 2024). Decision trees are one approach to

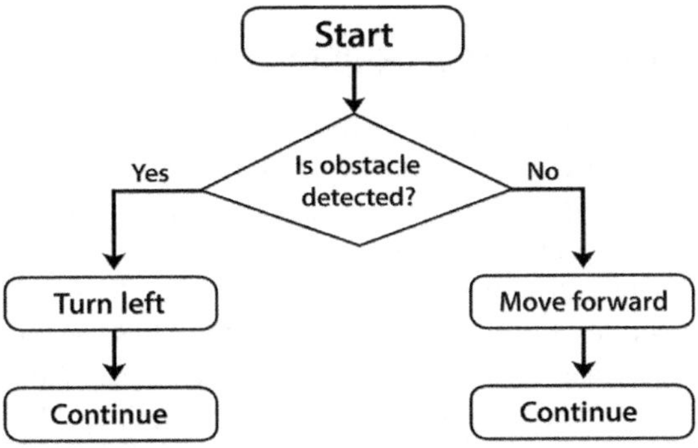

FIGURE 3.1 Decision tree model for intelligent systems.

systematising robotic decision-making and behaviour generation to ensure consistency and repeatability. They are popular in domains where interpretability and analysis of the decision-making process are still required.

3.4 CASE STUDY: AlphaZero BY GOOGLE DEEPMIND

AlphaZero uses deep neural networks in combination with Monte Carlo tree search (MCTS) as part of a self-play reinforcement learning algorithm. AlphaZero approaches games without using hardcoded rules to evaluate board positions. AlphaZero learns to evaluate chess positions by playing games against itself over time. AlphaZero learns to master games such as chess, Go, and shogi by learning from playing games, starting only from the rules of the game itself (Scheiermann & Konen, 2022).

Breaking down how AlphaZero works, AlphaZero selects moves and evaluates board positions using a learned policy–value network. The network then outputs a policy predicting the likely good actions and a value predicting the likelihood of winning from the current state (Trudeau & Bowling, 2023). Monte Carlo tree search is then used to expand upon thousands of possible move sequences using the current state as the root. AlphaZero uses what it has learned from its neural network to recognise patterns in each state it evaluates.

These topics closely relate to computational thinking. Decomposition is used by breaking down a game into a series of moves. AlphaZero uses pattern recognition by evaluating board positions and learning from those patterns over time (Schut et al., 2025). Finally, the system uses updated policy and value functions to iteratively learn from trials to improve.

AlphaZero's gameplay can be mapped to a decision tree, but each node wouldn't have a definitive child node. Instead, the tree would become incomprehensibly large, having one branch for every possible game state. Instead, AlphaZero uses MCTS and a neural network to predict which moves are better than others (Trudeau & Bowling, 2023). On that note, state is an important term to understand. By assigning a value to a state or board position, AlphaZero can make informed decisions in an exponentially large state space. In summary, AlphaZero can be thought of as an advanced set of if-else statements that do not follow a linear decision path and can generalise to any board state.

3.5 LOGIC AND CONTROL IN PROGRAMMING

Logical expressions can be used to control whether certain parts of code execute. This allows robots to determine when they should behave in certain ways or follow certain control policies. Robot controllers use logic statements to decide how to act given certain inputs (Cui et al., 2024). Logic statements are typically written using binary logic and control structures.

Binary logic operates on true and false values. These values map to Boolean conditions, which indicate facts about the state of the environment such as "sensor X detects an obstacle" or "task is finished" (Kurtz & Lin, 2022). Robot controllers use predicates (conditions that can be true or false) to represent constraints like limits on sensor values or requirements for the robot to finish a state (Yin et al., 2024).

Constraints are used to enforce robotic systems' safety requirements and could include predicates such as "object in region" (Kurtz & Lin, 2022).

Predicates can be combined using Boolean operators such as AND and OR. For example, safety constraints may be enforced using predicates combined with AND, while goals may be expressed using predicates combined with OR (Kurtz & Lin, 2022). The flexibility that comes with using logic allows for robotic controllers to be designed and formally verified (Wang & Chortos, 2022).

Logic can be implemented in code using control structures such as if–else statements, which specify what should be done in a certain condition. Control structures like loops can sense the state of the environment and act continuously, which allows for the creation of reactive behaviour and feedback control (Jordana et al., 2024).

Logic and control structures allow for programmable behaviour and decision making in robotics (Cui et al., 2024).

3.6 PRACTICAL INSIGHT: PROGRAMMING LOGICAL BEHAVIOUR

Decision trees are a way of representing decisions at a high level, but we must also encode the decisions into code. We can use programming languages to turn our logical models into control software that drives autonomous agents' behaviour (Cui et al., 2024; Yang & Zhang, 2023). Instead of using trees to represent decisions, programming languages can express decisions as if-then rules and control structures like loops and conditionals or more abstract representations of tasks like finite-state machines and behaviour trees (Ghzouli et al., 2022). Robots' behaviour can be specified at a high level in a way that is both machine-replicable and allows for modular, reactive behaviour, and formal verification of control behaviour (Domínguez et al., 2022).

Consider the rudimentary robot that will stop if it detects an obstacle in front of it. Obstacle avoidance is a typical problem in mobile robotics and navigation (Bernardo et al., 2023; Domínguez et al., 2022). The robot will continuously sense the world and reason about what action to take based on those sensor readings. This is commonly framed as a set of predicates over the objects detected in the world and their distances measured in a control loop (Cui et al., 2024; Yang & Mao, 2022). At its simplest, this behaviour can be programmed with straightforward control flow statements like if, Elif, and else or through a domain-specific language that makes condition–action rules and control loops available to write robot-control programs (Cui et al., 2024; Yang & Zhang, 2023).

The robot might check if there is an obstacle in front of it based on some sensed threshold. We can encode this decision as a Boolean condition in our control program (Bernardo et al., 2023; Yang & Mao, 2022). If there is an obstacle in front of the robot, the robot will then perform some obstacle avoidance behaviour like turning away or replanning its path with some behaviour tree or stack-of-tasks controller (Domínguez et al., 2022; Yang & Zhang, 2023). If not, it will simply proceed forward with the default policy of "go forward." This logic is equivalent to the decision tree shown previously, where each branch represents an if condition that leads to some outcome. Condition-action rules are commonly implemented as hierarchical behaviour trees or finite-state machines that translate discrete nodes into chunks of code (Ghzouli et al., 2022).

Listing 3.1: Conditional logic for obstacle avoidance in Python

```
if obstacle_detected:
turn_left()
elif obstacle_near:
slow_down()
else:
move_forward()
```

The above example showcases how conditional statements work at a high level: they provide a mapping from logical conditions to actionable instructions. This process is very common in programming languages used to control robots, with symbolic rules describing the task at hand being translated into movement instructions (Domínguez et al., 2022). By evaluating each statement from top to bottom until a condition has been met, the corresponding piece of code will be executed. The presented example implements what is known as rule-based (or policy-based) control: depending on which state the system is in, certain behaviours will be performed. States can be extracted from raw sensor data using perception pipelines, and behaviours can entail various movement commands. Using this type of software architecture allows for reasoning about the system's response to different situations, which enables guarantees about its safe operation, reliability, and verifiability (Wang & Chortos, 2022).

Loop constructs allow us to leverage the capabilities in repeatable ways, providing the basis for so-called perception–action loops used in traditional embedded controllers as well as learning-based systems (Chen & Chung, 2023). Robots must continuously perceive their environment using sensors and make decisions on-the-fly to account for changes in the environment and the task itself. By placing decision-making logic inside loops, the system can re-evaluate conditions at every time step and adapt its behaviour accordingly. This allows for implementing control strategies for tasks such as obstacle avoidance, trajectory tracking, and human–robot interaction. With implementations of logic and decision-making, we are one step closer to harnessing the power of these abstractions to control our technological creations.

3.7 DECOMPOSING ROBOTIC BEHAVIOUR

Decomposition encourages students to think about systems hierarchically, breaking complex problems into simpler modules or behaviours and understanding how behaviours manifest at different system levels (Chichekian et al., 2023). Rather than considering a behaviour holistically, decomposition asks students to break down the constituent pieces that make up a behaviour, then focus on implementing and testing those pieces individually before integrating them into the whole system (Ghzouli et al., 2022; Yang & Zhang, 2023). In robotics, missions can be decomposed into granular skills, as well as high-level coordination systems like behaviour trees or finite state machines (Yang & Zhang, 2023).

Robotic systems can feel daunting or overly complex because they tend to execute multiple processes at once and operate at runtime in unpredictable and changing environments (Yang & Zhang, 2023). Avoiding obstacles, for instance, requires acting, sensing, perception, and planning. Through decomposition, this task can be broken down into logical steps: sense obstacle, determine distance, choose new path, then act (Allard et al., 2022). Each step can be mapped to reusable modules or algorithms used in a robot's control schema (Ghzouli et al., 2022; Yang & Zhang, 2023).

Programming a robot often uses many of these same design concepts. Robots exhibit complex behaviours by coupling functions and calling upon disparate but well-defined modules to sense, plan, and act on their environment. Decomposition encourages students to think about these systems as separable and isolated, which can make them more manageable to develop and maintain. Modules can be swapped out or upgraded independently of the rest of the system (Ghzouli et al., 2022).

The previously mentioned concepts fit hand-in-hand with decomposition as well. By thinking of robot behaviours as smaller actionable steps, students can more easily reason about why a system is taking specific actions. It also allows for smaller chunks of the overall system to be refined or troubleshooted independently of the rest (Allard et al., 2022; Chichekian et al., 2023). Understanding how complex behaviours are created from smaller logical parts is beneficial to learn for students (Chichekian et al., 2023).

3.8 CHALLENGES IN COMPUTATIONAL THINKING

Computational thinking enables reasoning. Logic-based/Symbolic AI will often need to be translated into a form that allows computational reasoning about a problem ultimately understandable by a human. This usually means that some finite set of logical rules must be followed to arrive at some decision by a machine or that an objective function needs to be optimised (Zeng et al., 2023). Reasoning about the real world is more abstract than this; however, worlds can often be ambiguous and ill-defined and change over time. Logic requires things to be clearly defined; however, humans reason about goals, context and intuition vs. conditions and constraints.

The inherent complexity of most robotics' problems present in the real world makes them difficult to decompose. Many real-world robotics tasks consist of perception, planning, control, and actuation, all coupled together over time in a changing environment. Computational thinking promotes decomposition of a problem into its parts. But in robotics, these parts all depend on each other, and so decomposing the system will often require coupling it back together in some way. Therefore, not only is decomposition difficult, but it can often lead to more complexity.

Rule-based systems/systems defined by logic are easy to reason about when you have a small number of rules/features. As the complexity of the system grows, so does the number of states it can take on. Large numbers of rules/features can lead to combinatorial explosion, making the system hard to interpret. For this reason, many real-world systems move away from these kinds of logical representations to other methods like machine learning that lack interpretability but scale better (Zeng et al., 2023).

Uncertainty is also difficult to represent with computation/logic. Robotics systems will often have to reason about uncertainty from noisy sensors, their environment, and partial observability of that environment. Logical representations often have to be augmented with probabilities to reason about this uncertainty, making systems less interpretable but more robust to noise (Zeng et al., 2023).

3.9 ETHICAL AND REAL-WORLD CONSIDERATIONS

Computational thinking for intelligent systems poses several limitations and challenges that should be considered.

3.9.1 Ethical and Social Considerations

When systems make decisions without human oversight, there may be ethical and social consequences. Bias mitigation and responsible AI should be considered when designing high-stakes systems that may produce unfair or harmful outcomes (Cheong, 2024). Logic alone may not be enough to solve complex problems

3.9.2 Bias in Decision-Making

Systems that follow pre-programmed rules or are trained on historical data may replicate and amplify existing biases, coding mistakes, or assumptions that are present in the data they use (Pagano et al., 2023). If rules are biased or incomplete, systems can exhibit poor performance on edge cases and harm particular populations. This can be problematic in scenarios where decisions have a significant impact on people's lives, such as healthcare, finance, or criminal justice systems (Zafar, 2024).

3.9.3 Limited Flexibility and Generalisability

Logical models require clearly defined inputs and thresholds, which can make them brittle and less able to handle unexpected situations or adapt to changing environments (Nishant et al., 2023). Machines may also lack the "common sense" that humans apply to decision-making, limiting their flexibility in certain contexts. For example, while self-driving cars may know the rules of the road, they may still struggle to navigate ambiguous situations that a human driver would instinctively understand (Tiwari & Farag, 2024).

3.9.4 Lack of Transparency

While rules and decision trees are inherently interpretable, many machine learning models are considered "black boxes" that even their creators cannot understand. It can be hard to know why an AI system made the decision that it did (Singhal et al., 2024). Transparency becomes problematic in high-stakes applications where decisions need to be explainable (Cheong, 2024).

3.9.5 Responsibility/Accountability

When automated systems malfunction, it may not always be clear who is responsible for their decisions. Ensuring responsibility requires clear governance, monitoring, and human oversight. Implementing principles of responsible AI can help ensure your system is safe and equitable. Responsible AI includes fairness, accountability, transparency, and ethics (FATE) among other things (Pagano et al., 2023).

3.10 CONCLUSION

The key topics and concepts covered in this chapter included computational thinking and how problem-solving can be formulated into logical steps that can be understood by computers. Decomposition, pattern recognition, abstraction, and algorithms allow real-world issues to be broken down into sets of instructions that a computer can follow. Concepts such as decision trees provided structure to allow visualisation of how a system can take stock of its current conditions and pick a course of action. Google DeepMind' AlphaZero showed how these concepts can be used in complex systems that are also able to recognise patterns and learn from them. Condition statements and loops were introduced showing students the connection between logical models and how these are represented in programming languages. Sample challenges allowed students to see how rules and behaviour of robots can be translated into lines of code. Topics such as challenges and ethical concerns were briefly covered to show some of the downsides of rule-based programming and allow students to think about how they design their own systems. These ideas will help further develop students' knowledge as they progress to the next chapter on programming techniques and implementing these on systems.

3.11 DISCUSSION QUESTIONS

1. What is computational thinking? Why is it important for intelligent systems?
2. How do decomposition and abstraction help us solve problems with robots?
3. Describe how decision trees can be used to model the decision-making process of an intelligent system.
4. Describe ways in which AlphaZero used computational thinking.
5. How do computers "see"? How do they solve problems?

3.12 EXERCISES

Exercise 1: Build-a-Bot Decision Tree

Create a decision tree for a bot. This could be something as simple as navigating through a room with obstacles, following a line, etc. Be sure to define what conditions your bot will use to make decisions, as well as what actions it will take.

Exercise 2: Writing Pseudocode

Write some pseudocode that allows a robot to navigate through a room while avoiding obstacles. Remember to use condition statements to break up the steps.

Exercise 3: Translating Logic into Code

Take a simple set of decisions that you would like to program and convert it to Python-style logic code using if–elif–else statements. Describe how each decision affects what the robot will do.

Exercise 4: Reflection on Computational Thinking

Compare and contrast how humans and computers solve problems. What are some pros and cons of each?

REFERENCES

Allard, M., Smith, S. C., Chatzilygeroudis, K., Lim, B., & Cully, A. (2022). Online damage recovery for physical robots with hierarchical quality-diversity. *ACM Transactions on Evolutionary Learning*, *2*(4), 1–28.

Amri, S., Budiyanto, C., Fenyvesi, K., Yuana, R., & Widiastuti, I. (2022). Educational robotics: Evaluating the role of computational thinking in attaining 21st century skills. *Open Education Studies*, *4*(1), 219–241.

Barradas, R., Lencastre, J. A., Soares, S. P., & Valente, A. (2024). Arduino-based mobile robotics for fostering computational thinking development: An empirical study with elementary school students using problem-based learning across Europe. *Robotics*, *13*(11), 159.

Bernardo, R., Sousa, J. M. C., Botto, M., & Gonçalves, P. J. S. (2023). A novel control architecture based on behavior trees for an omni-directional mobile robot. *Robotics*, *12*(1), 10.

Chen, C.-H., & Chung, H.-Y. (2023). Fostering computational thinking and problem-solving in programming: Integrating concept maps into robot block-based programming. *Journal of Educational Computing Research*, *61*(8), 1593–1619.

Cheong, B. C. (2024). Transparency and accountability in AI systems: Safeguarding well-being in the age of algorithmic decision-making. *Frontiers in Human Dynamics*, *6*, 1405673.

Chichekian, T., Trudeau, J., Jawhar, T., & Corliss, D. (2023). Experimenting with computational thinking for knowledge transfer in engineering robotics. *Journal of Computer Assisted Learning*, *39*(6), 1570–1588.

Cui, G., Wang, Y., Qiu, W., & Zhu, H. (2024). Reward-guided synthesis of intelligent agents with control structures. *Proceedings of the ACM on Programming Languages*, *8*(POPL), 1–30.

Domínguez, D., Iannotta, M., Stork, J. A., Schaffernicht, E., & Stoyanov, T. (2022). A Stack-of-Tasks approach combined with behavior trees: A new framework for robot control. *IEEE Robotics and Automation Letters*, *7*(4), 10847–10854.

Ertas Kalay, S., & Ipek, J. (2024). Bibliometric analysis of artificial intelligence studies in computational thinking. *International Journal on Studies in Education*, *6*(4), 1–24.

Ghzouli, R., Berger, T., Johnsen, E. B., Wąsowski, A., & Dragule, S. (2022). Behavior trees and state machines in robotics applications. *IEEE Transactions on Software Engineering*, *48*(12), 4988–5009.

Huang, X. (2024). Predictive models: Regression, decision trees, and clustering. *Applied and Computational Engineering*, *4*(2), 45–63.

Jawawi, D. D., Jamal, N. N., Abdul Halim, S., Sa'adon, N. H., Mamat, R., Isa, M. A., Mohamad, R., & Abdull Hamed, H. N. (2022). Nurturing secondary school student computational thinking through educational robotics. *International Journal of Emerging Technologies in Learning*, *17*(4), 4–21.

Jordana, A., Kleff, S., Meduri, A., Carpentier, J., Mansard, N., & Righetti, L. (2024). Structure-exploiting sequential quadratic programming for model-predictive control. *IEEE Transactions on Robotics*, *40*(1), 1–16.

Kurtz, V., & Lin, H. (2022). Mixed-integer programming for signal temporal logic with fewer binary variables. *IEEE Control Systems Letters*, *6*, 2602–2607.

Nishant, R., Schneckenberg, D., & Ravishankar, M. N. (2023). The formal rationality of artificial intelligence-based algorithms and the problem of bias. *Journal of Information Technology, 38*(3), 303–324.

Pagano, T. P., Loureiro, R. B., Lisboa, F. V., Peixoto, R. M., Guimarães, G. A. S., Cruz, G. O. R., Araujo, M. M., Santos, L. L., Cruz, M. A. S., Oliveira, E. L., Winkler, I., & Nascimento, E. S. (2023). Bias and unfairness in machine learning models: A systematic review on datasets, tools, fairness metrics, and identification and mitigation methods. *Big Data and Cognitive Computing, 7*(1), 10.

Pirzado, F., Ahmed, A., Hussain, S., Ibarra-Vázquez, G., & Terashima-Marín, H. (2025). Assessing computational thinking in engineering and computer science students: A multi-method approach. *Education Sciences, 15*(3), 345.

Sarker, I. H. (2022). AI-based modeling: Techniques, applications and research issues towards automation, intelligent and smart systems. *SN Computer Science, 3*(3), 185.

Scheiermann, J., & Konen, W. (2022). AlphaZero-inspired game learning: Faster training by using MCTS only at test time. *IEEE Transactions on Games, 14*(4), 605–617.

Schut, L., Tomašev, N., McGrath, T., Hassabis, D., Paquet, U., & Kim, B. (2025). Bridging the human–AI knowledge gap through concept discovery and transfer in AlphaZero. *Proceedings of the National Academy of Sciences, 122*(10), e2311234122.

Singhal, A., Neveditsin, N., Tanveer, H., & Mago, V. K. (2024). Toward fairness, accountability, transparency, and ethics in AI for social media and health care: Scoping review. *JMIR Medical Informatics, 12*, e51865.

Sirichanya, C., & Kesorn, K. (2023). Semantic decision trees: A new learning system for the ID3-based algorithm using a knowledge base. *Advanced Engineering Informatics, 57*, 102054.

Tiwari, A., & Farag, H. E. Z. (2024). Responsible AI framework for autonomous vehicles: Addressing bias and fairness risks. *IEEE Access, 12*, 78291–78315.

Trudeau, A., & Bowling, M. H. (2023). Targeted search control in AlphaZero for effective policy improvement. In *Proceedings of the 22nd International Conference on Autonomous Agents and Multiagent Systems* (pp. 1530–1539).

Wang, J., & Chortos, A. (2022). Control strategies for soft robot systems. *Advanced Intelligent Systems, 4*(6), 2100162.

Wu, T.-T., Asmara, A., Huang, Y.-M., & Hapsari, I. P. (2024). Identification of problem-solving techniques in computational thinking studies: Systematic literature review. *SAGE Open, 14*(2), 1–17.

Yang, S., & Mao, X. (2022). Towards a behavior tree-based robotic software architecture with adjoint observation schemes for robotic software development. *Automated Software Engineering, 29*(2), 17.

Yang, S., & Zhang, Q. (2023). Towards efficient robotic software development by reusing behavior tree structures for task planning paradigms. *Complex Systems Modeling and Simulation, 3*(4), 263–285.

Yu, D., Wang, H., Cao, X., Wang, Z., Ren, J., & Zhang, K. (2024). Enhancing autonomous underwater vehicle decision making through intelligent task planning and behavior tree optimization. *Journal of Marine Science and Engineering, 12*(5), 879.

Yuan, J., Barr, B., Overton, K., & Bertini, E. (2022). Visual exploration of machine learning model behavior with hierarchical surrogate rule sets. *IEEE Transactions on Visualization and Computer Graphics, 28*(12), 4771–4786.

Zafar, A. (2024). Balancing the scale: Navigating ethical and practical challenges of artificial intelligence (AI) integration in legal practices. *Discover Artificial Intelligence, 4*(1), 23.

Zeng, Z., Cheng, Q., & Si, Y. (2023). Logical rule-based knowledge graph reasoning: A comprehensive survey. *Mathematics, 11*(21), 4429.

Part II

Programming and Software Systems

4 Programming Smart Agents

4.1 INTRODUCTION

Programming can be thought of as turning our perception, decisions, and actions into code that we unify under software architecture (Milojevic et al., 2025). Programs give embodied intelligence, math models, and computational thinking tasks the ability to run perception, control, and planning for actual systems (González-Santamarta et al., 2024). Meaning we can take theoretical code and place it into the real world to make decisions (Mittal et al., 2023).

Robotics programming dictates how an autonomous system behaves in relation to its sensor perception through repeated loops of perception–decision–action (Milojevic et al., 2025). Robots read inputs from sensors, make decisions through logic or learned processes, and send commands to various actuators. These tasks can be organised through task graphs, behaviour trees, or learned policy meaningfully, creating the "brains" of our autonomous agents (Kawaharazuka et al., 2025).

Python is commonly used in robotics because it's readable, versatile, and has many scientific and machine learning libraries (Vemprala et al., 2023). It is very beginner-friendly, so users can learn about programming without getting lost but can still program for robust real-time robotic applications.

4.2 FOUNDATIONS OF PROGRAMMING FOR INTELLIGENT SYSTEMS

Programming concepts allow one to codify logical reasoning processes that map input to output and create logic for decision-making; typically called perception–decision–action loops (Programming1). Programming languages at the lowest level tell the system how it should behave when presented with specific information. Variables, data types, behaviours, state machines, control flow, input/output are elements in programming that allow us to build robotic systems that sense and can operate in the world. State machines or behaviour trees are widely used abstractions of behaviour that allow complex intelligent behaviour to be represented in hierarchical structures for readability and modularity (Ghzouli et al., 2022).

Variables are used to store data that may change during program execution. This memory typically represents information about the system itself that can be used for robot control or planning purposes (Chen et al., 2023). Variables can often be mapped to sensor measurements or system states. In a feedback loop, these may represent set points for control or a sensor that helps dictate decision-making (Wang & Chortos, 2022). Position, velocity, distance to an obstacle, and motor speed are all examples of variables that a robot may use to reason about the world (Meli et al., 2023).

DOI: 10.1201/9781003758501-6

Data types are intrinsically linked to variables. Variables have data types that allow them to be manipulated during program execution. Integers and floating-point numbers are used for mathematical computation. Booleans allow for reasoning about conditions and are often used in control flow structures (Chen et al., 2023; Wang & Chortos, 2022). Input and output can be thought of as how information enters the system and how it leaves. Robots operate in the world by using sensors to perceive their environment and by using actuators to output into the environment (Meli et al., 2023). Robots can also represent their state through position or sensor values.

4.3 STRUCTURING BEHAVIOUR IN CODE

Programming requires writing code in logical instructions for a robot to take input, process it, and produce output intelligently. In closed-loop, perception–decision–action pipelines intelligent systems (Kondratenko et al., 2022; Wang & Chortos, 2022) process inputs to make logical decisions. Outputs follow completing the process instruction. Programming thus means defining how an AI system acts based on its input by breaking down abstract decision-making into executable instructions that a machine can follow. Logic instructions are sometimes represented as state machines or behaviour trees that program complex behaviours through multiple sub-actions and control loops.

Inputs are ways in which the robot perceives its environment and itself. Robots take inputs from sensors related to the space around them and of themselves. Distance, light, proprioceptive sensors, and vision or touch are common input types that allow a robot to sense its position and surroundings (Hegde et al., 2023). Internal states like motor speed or force applied on a joint are other input types. Inputs must be updated regularly to allow controllers to act and react to their environment and task.

Processing includes any instruction that makes logical sense of the input. Implementations of control can follow rule-based logic, model-based, and learning-based algorithms (Wang & Chortos, 2022). Processing includes if-else conditions, calculations, and motion instructions like how to move forward, turn, or stop. Code that implements computation to make decisions can take the form of if-else statements, loops, and code blocks. Higher-level computation can be performed by task planners and can be abstracted as behaviour trees (Ghzouli et al., 2022).

Outputs are any action that the robot executes. This is typically done through actuators that move the robot or parts of the robot. Movements can include traversing in the environment or changing the speed, deformation, or applied force. Outputs close the loop by responding to what is sensed by the robot. Robots commonly use feedback from sensors to act and react to the environment to perform tasks reliably (Kondratenko et al., 2022; Mittal et al., 2023).

Figure 4.1 depicts this organisation as a cycle. Environmental input is filtered through decision logic. The actions resulting from that logic affect the environment. In turn, those actions produce more input from the environment. This model closely resembles the cycle of intelligent behaviour introduced in previous chapters.

Programming behaviour like this makes programs easier to design, understand, and maintain. It forces you to cleanly separate sensing, deciding, and doing. Each part of the system can do its job without confusion.

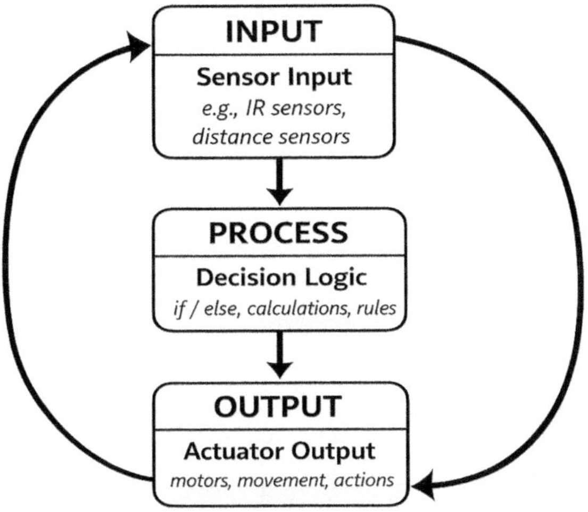

FIGURE 4.1 Program flow model for intelligent agents.

4.4 CASE STUDY: LINE-FOLLOWING ROBOT IN PYTHON

Programming serves as the glue between sensing, decision-making, and acting to produce intelligent robot behaviour (Farkh & Aljaloud, 2022; Leal et al., 2025). For example, line-following robots have controlled movement along a visible path by processing sensor input with decision logic. Arduino and Raspberry Pi robotic platforms reveal perception, planning, and action coupled together through program execution (Leal et al., 2025; Somesh & Aruna, 2025).

IR sensors mounted underneath are used at the sensing layer to observe the line relative to the chassis. Some robots use cameras and apply computer vision methods instead. Sensors sample the world at fixed frequencies, producing values that quantify how far off track the robot is (Farkh & Aljaloud, 2022). Sensor values must be handled rapidly enough to allow for control signals to react to disturbances or tracking errors.

At the decision-making layer, logic is applied to sensor values to select an action. Often this behaviour is hard coded as conditional branches, like turn left if the line is left of centre or turn right if the line is right of centre. Control theory can also be applied to dictate actions to achieve smooth tracking with minimal error (Farkh & Aljaloud, 2022; Leal et al., 2025). These types of conditional behaviours can be implemented as if–elif–else constructs or rules (Somesh & Aruna, 2025).

The selected action is performed at the output layer by commanding the motors to move. Motor control values often translate to wheel speeds that use pulse-width modulation or drive kinematics such as differential drive (Farkh & Aljaloud, 2022; Somesh & Aruna, 2025).

Each of these layer's repeats in a loop allowing for feedback control. The perception–decision–action schema captures important aspects of the control strategies used in autonomous robots today (Leal et al., 2025).

Listing 4.1: Line-following control logic in Python

```
left_sensor = read_left_sensor()
right_sensor = read_right_sensor()

if left_sensor and right_sensor:
move_forward()
elif left_sensor:
turn_left()
elif right_sensor:
turn_right()
else:
stop()
```

This demonstrates how sensor information becomes movement through logic gates. Each logical statement is very clearly defined so you know exactly how the robot will act in any situation.

It really shows how programming is the medium through which intelligence is implemented. Combining sensor data with some logical if statements and outputting to actuators can allow even simple robots to act in smart ways. It also shows you should always write nice modular code and test it so you can make your robots faster and smarter.

4.5 CORE PYTHON CONCEPTS FOR ROBOTICS

Variables and data structures provide programming languages with means to store the internal state of the program, such as sensor readings, actuator values, task parameters, etc. (Patton et al., 2023). Variables are used in robotics to represent sensed data such as colours, distances, or motor speeds (Kondratenko et al., 2022; Wang & Chortos, 2022). They can also represent more abstract concepts, such as properties of the environment or parameters for how a robot/system should behave (Vemprala et al., 2023).

Conditionals give programs the ability to decide between different actions depending on whether some statement is true or false. Robot controllers can be programmed to make decisions based on sensor values using conditionals. For example, if something is sensed then turn otherwise move forward (Patton et al., 2023). Conditionals form the basis of reactive control policies and are present in most task-level languages and behaviour trees. Even learned controllers such as neural networks often contain structured conditional statements (Ghzouli et al., 2022; Patton et al., 2023).

Loops are constructs in programming languages which allow a program to repeat an instruction or a collection of instructions. In robotics, loops can be used to repeatedly sense, decide and actuate, forming a control loop. This allows robots to operate in real-time as they sense their environment and make decisions based on sensor readings. Loops can be used to structure periodic operations but also event-driven loops, e.g., to construct a navigator or enable long autonomy horizons (Kondratenko et al., 2022; Mittal et al., 2023; Patton et al., 2023).

Functions allow programmers to abstract away specific behaviour or capabilities into blocks of code. Sensing, deciding, and actuating could all be implemented as individual functions, for example. This makes code more readable and allows smaller, more manageable programs to be constructed. Functions allow for hierarchical decomposition of tasks into smaller subprograms. Most modern robotic frameworks, as well as program synthesis tools, utilise functions as a fundamental building block of programs (Mittal et al., 2023; Patton et al., 2023; Vemprala et al., 2023).

4.6 WORKING WITH SENSOR INPUT

Sensor input is information read from a robot's sensors about the world and received by a program. It is then used by the program to perceive its environment, make decisions about what it should do, and control actuators (Hegde et al., 2023; Kondratenko et al., 2022). From a programming perspective, sensor input involves reading values from sensors, assigning meaning to those values, and mapping those sensor readings to actions using control laws or other signal processing techniques (Ji et al., 2022). Sensor input allows robots to dynamically interact with their environment and adapt their behaviour based on external stimuli while remaining robust to noise and uncertainty.

Signals from sensors represent aspects of the world. For example, a sensor can provide a lidar point cloud, an image from a camera, an infrared reading, or information about whether the robot is touching something with its whiskers or hands (Hegde et al., 2023). In a program, sensor inputs can be stored as numeric or Boolean values. For instance, consider a line-following robot that uses infrared sensors to see if it is on the line (Kondratenko et al., 2022). The robot's infrared sensors might produce values that can be stored as on-line and on-right variables to help the robot follow lines.

Thresholds, rules, or probabilities may then be used to interpret the sensor inputs to allow decision-making. For example, we may decide that an object is "near" or "far" from the robot based on the input value. Sensor inputs can be used with fuzzy logic to manage imprecise information (Ji et al., 2022). When programming robots or other autonomous systems, inputs can indicate the presence of obstacles, identify features, and activate the correct behaviours with if statements, control policies, etc. (Karle et al., 2023).

Sensor input values are often read continuously as part of a control loop. Because of this, many robots can sense their environment and update their actions quickly and automatically. Sensor noise and errors can occur while reading sensors and dealing with the resulting data, so it is essential to filter, fuse, and validate sensor inputs (Maken et al., 2025).

4.7 DATA STRUCTURES IN ROBOTIC SYSTEMS

Data structures can help programmers organise information used by the robot. They enable groups of values, or more complicated data formats, to be stored, accessed and processed efficiently throughout program execution. Sensor streams, system states, and actions can all be encoded using data structures; all these inputs inform decision-making in robotics (Hughes et al., 2024). Many modern robotics programs are written to incorporate structured data such as lists, maps and trees to represent inputs, outputs, and internal program representations (Macenski et al., 2022).

Lists (arrays) store a sequence of values that typically relate to each other. A robot may have multiple sensors providing distance values that should be stored somewhere in memory. By placing these distance values into a list, one can easily program the robot to iterate through each element and compare these values. Lists are useful to process bulk inputs together or when ordering is important, such as checking for obstacles on either side of a line follower robot (Mittal et al., 2023).

Dictionaries organise data into key-value pairs. Inputs can be mapped to their corresponding actions or behaviours using intuitive keys. Sensor readings and symbolic labels can both serve as effective dictionary keys to output intelligible behaviours like movement controls. Dictionaries offer readable and flexible code and can be useful to organise data for use in tasks such as planning or semantic mapping (Ghzouli et al., 2022; Hughes et al., 2024; Kawaharazuka et al., 2025).

States or data about the robot itself can also be stored using data structures. A robot's position, sensor history or even operating mode can be stored as values to allow the system to reason about itself. Such information allows robots to use both past and present information to complete advanced tasks like localisation and sensor fusion (Hughes et al., 2024).

Programming with structured data allows for extensible and modular program design. Many modern robotics software architectures share this philosophy, allowing for additional sensors or subsystems to be added by including new elements of a data structure rather than rewriting code (Macenski et al., 2022).

Fundamentally, by allowing for information to be processed quickly and efficiently, data structures enable intelligent behaviours in robots (Kawaharazuka et al., 2025; Mittal et al., 2023).

4.8 PRACTICAL INSIGHT: BUILDING A LINE-FOLLOWING ALGORITHM

Building a line following algorithm is beneficial since it will help students visualise how code is used to develop intelligent behaviour in robots and how sensing connects to control logic to action in a feedback loop (Farkh & Aljaloud, 2022; Leal et al., 2025). A robust line-following algorithm makes decisions based on sensor readings and associated actions, completing the control loop. It repeats the reading-evaluating-actuating pattern in an endless loop with an embedded controller and an encoded feedback control law, allowing it to follow a line in real-time (Baluja, 2024).

Programming a line following algorithm typically starts with determining how to read from sensors. A line-following robot will commonly have infrared sensors reading the reflectance differences of the line and pavement or colour sensors/cameras that identify the line's colour and geometry (Farkh & Aljaloud, 2022). Mapping how far away from the line the robot can be done by reading if the line sensors are under the line or not (Baluja, 2024). Three classifications can be made regarding whether the line is detected by the left sensor, right sensor, or both sensors. For instance, if both sensors read that they are on the line, the robot knows it is going straight.

The second step is to define the if-else conditions. This can either be done by hard-coding rule-based logic or creating PID-based actions to track the line. If only the left sensor reads that it is on the line, an action can be taken to move the robot slightly left. If only the right sensor reads that it is on the line, then an action can be taken to move the robot slightly to the right. If both sensors read that they are on the line, then move forward. As demonstrated here by these examples, if-else statements can create associations between sensor readings and actions. If the robot detects that it is off the line with both sensors, it can either stop or enter a recovery routine to get back on the line. Some algorithms will learn the proper policies to execute when off track based on training (Leal et al., 2025). Actions can thus be created for any combination of sensor readings (Baluja, 2024).

The last step is to wrap the code in a loop. Placing the code within a loop allows the robot to read sensors, update conditions, and update movements continually. By repeating these steps, the code will operate in real-time using a feedback control methodology. Following these steps will allow the robot to follow a line by reacting to deviations from the line and recovering from mistakes, allowing for stable line tracking with minor inconsistencies in the line.

The following listing shows a stripped-down version of the algorithm in Python.

Listing 4.2: Line-following algorithm implemented in Python

```
while True:
    left_sensor = read_left_sensor()
    right_sensor = read_right_sensor()

    if left_sensor and right_sensor:
        move_forward()
    elif left_sensor:
        turn_left()
    elif right_sensor:
        turn_right()
    else:
        stop()
```

This trivial example shows sensor reading, decision-making, and behaviour producing combined in feedback cycle. Due to the cyclic nature of this program, the robot is able to dynamically adjust its behaviour to the environment.

The modularity of this design allows functions like read_left_sensor() or move_forward() to describe behaviour by themselves. Modular code makes it easier to read, test, and change behaviour. Modularity also allows your robot's behaviour to grow more complex over time.

Developing and stepping through these kinds of algorithms allow students to visualise how code translates into responsive, smart behaviour.

4.9 DEBUGGING AND CODE RELIABILITY

Debugging the logical aspects of code include checking that the program behaves as intended and your robots execute its behaviour correctly (Araujo et al., 2022; Caldas et al., 2024). Logical bugs in software can result in your robots doing unexpected or wrong behaviours. For complex intelligent systems, it is important to debug correctly.

Safety-critical and autonomous robots often can't prove their correctness at design time and have the potential to harm the real world if bugs exist.

Wrong robot behaviours can easily occur if your conditional statements or behaviour requirements are off by a small amount. For example, if part of your robot's task constraints are violated because you accidentally programmed your robot to do the opposite of what is needed (Rizwan et al., 2025). Checking your if statements and ensuring your finite-state machines or behaviour trees make logical sense can help spot these errors.

False readings from sensors can often cause your robot to behave strangely or even unsafely. Robot sensors can often provide noisy or inaccurate data. Debugging these issues often comes down to checking that the sensors and readings you expect are correct. Filtering/sensor fusion and checking that your perception pipelines are working as intended by testing will help (Araujo et al., 2022).

Organising your code in a clean readable way will help you debug whether you're programming a single robot or distributed architectures like ROS2 (Macenski et al., 2022). Bugs are often caused by a lack of modularity in your code. Making use of functions and clearly defined components can help you find bugs and keep your code organised (Chen, 2025).

Other helpful debugging techniques include unit testing your code, adding lots of print statements/logging, using tracers, and developing your code incrementally by testing along the way (Caldas et al., 2024).

4.10 CHALLENGES IN PROGRAMMING INTELLIGENT SYSTEMS

Programming for autonomous and intelligent systems is hard since the software needs to closely interact with the physical world and numerous robotic subsystems. Robots need to function in real-time, manage uncertainty, and integrate sensing, planning, and actuation seamlessly (Milojevic et al., 2025).

Timing constraints are one problem domain. Robot controllers must adhere to hard timing constraints since they must acquire data from sensors and issue actuation commands within a predetermined time frame. Control systems that operate late or miss deadlines risk becoming unstable or behaving dangerously, particularly during operations like collision prevention or cooperative motion. The underlying software and middleware layers must be adaptable enough to provide predictable, low-latency operations to meet timing constraints (Foughali & Zuepke, 2022; Macenski et al., 2022).

Hardware–software coadaptation is another issue. Physical constraints, including limited processing power, battery capacity, network bandwidth, sensing range, and so on, must all be considered when developing control algorithms. Performance may degrade due to inaccurate modelling of hardware components, which can result in incorrect system decisions. Efficient hardware–software co-design methodologies and practices must be created as a result (Foughali & Zuepke, 2022; Milojevic et al., 2025).

Uncertainty and robustness are other difficulties. Robots function in noisy environments with moving things, changing lighting, and uncertain hazards such as potholes and people. Robot control systems must be robust to fluctuations in the environment, while uncertainty-reduction actuators like sensors may increase robot robustness. Too many adaptable features, on the other hand, might cause the code to become bloated and unmanageable. Robot systems will almost certainly become more complex as they continue to include more features and adaptability. Managing software variability, implementing modular architectures, and model-based design are some options for dealing with these difficulties (Macenski et al., 2022).

The robot must finally scale to complete more difficult duties and missions while becoming more trustworthy as it learns how to handle simple chores. Layered software architectures, digital twins, adaptive/self-adaptive techniques, and feature extraction must all assist with scalability while maintaining a balance between performance, versatility, and safety (Macenski et al., 2022).

4.11 ETHICAL AND REAL-WORLD CONSIDERATIONS

Robotics requires careful consideration from an ethics and responsibility standpoint. Robots must work in environments built for people, governed by our expectations, safety standards, and regulatory frameworks. Therefore, engineering performance is only one aspect of robotics (Díaz Rodríguez et al., 2023). Ethical design and responsible research and innovation practices are prerequisites for safe deployment in safety-critical fields since robots should operate safely, reliably, and predictably in unstructured environments with uncertainties (Foughali & Zuepke, 2022). Much of our trust in these systems is placed on whether the code they use to function is correct and makes the right decisions.

Responsibility can also relate to system reliability and the robot's ability to sense its environment safely. Incorrect coding and unanticipated actions can arise from badly tuned learned policies, causing erratic actions, mission failures, or even dangerous behaviour (Foughali & Zuepke, 2022). Damaged or noisy sensors can also cause a robot to perceive its environment incorrectly, leading to accidents during locomotion or interaction with people (Ji et al., 2022). Verification, monitoring, and sensing validation are critical components to ensuring these systems operate correctly.

Robotic systems should also be transparent. We should know how and why the robots we interact with make the decisions they do. Code readability, explainability techniques, and robust accountability measures allow for safer, more trustworthy robotics that we can responsibly interact with in society.

4.12 CONCLUSION

This chapter introduced programming as the means to realising intelligent robotic behaviour. You learned how concepts presented throughout previous chapters come together as implementable code, allowing for real-time environmental interaction. Fundamental programming concepts, including variables, control structures, loops, functions, and data structures, were discussed as building blocks for creating structured and modular programs. The concepts learned throughout this chapter were exemplified by way of a line following robot by integrating sensor data acquisition, decision-making logic, and actuator control to create responsive behaviour.

Additional practical considerations were made for developing algorithms and debugging code to allow for the creation of clear, reliable, and testable programs. Challenges discussed highlighted the difficulties present when working with real-time systems, as well as the various considerations that must be made when working with hardware and accounting for real-world environmental conditions. Ethical considerations were presented to remind you of the responsibility that comes with creating autonomous systems.

The concepts presented in this chapter serve as an introduction to programming intelligent agents. You are now prepared to learn about integrating these systems, allowing for communication between components, and creating more complex autonomous behaviour.

4.13 DISCUSSION QUESTIONS

1. Why do you think programming is referred to as the "brain" of an intelligent robotic system?
2. How do variables, conditionals, and loops allow us to create robotic behaviour?
3. Why are functions and modularity important when designing reliable systems?
4. In what ways can an error in your code lead to unexpected physical robot behaviour?
5. What occurs when your robot incorrectly interprets its environment? How can you fix this?

4.14 EXERCISES

Exercise 1: Writing Basic Robot Logic

Write a Python program using conditional statements to simulate robot behaviour based on some sensor input.

Exercise 2: Modifying a Line-Following Algorithm

Modify the following line following algorithm. Add in another conditional statement to slow down when a robot is approaching a curve.

Exercise 3: Debugging Challenge

Something is wrong with a line following robot's code. The robot does not follow line correctly. There could be many errors in the conditionals. What are some things you would check?

Exercise 4: Reflection on System Behaviour

Describe how programming can affect the behaviour of a system. Provide an example of how you may make an incorrect assumption in your code that causes unexpected behaviour.

REFERENCES

Araujo, H. L. S., Mousavi, M., & Varshosaz, M. (2022). Testing, validation, and verification of robotic and autonomous systems: A systematic review. *ACM Transactions on Software Engineering and Methodology*, *31*(4), 1–76.

Baluja, A. (2024). Design and development of a line following robot using the proportional-integral-derivative (PID) control system. *International Journal for Multidisciplinary Research*, *6*(12), 112–125.

Caldas, R. D., Pinera García, J. A., Schiopu, M., Pelliccione, P., Rodrigues, G., & Berger, T. (2024). Runtime verification and field-based testing for ROS-based robotic systems. *IEEE Transactions on Software Engineering*. *50*, 2544–2567.

Chen, J. (2025). Research on improvement of intelligent robot control system. In *2025 IEEE 8th International Conference on Information Systems and Computer Aided Education (ICISCAE)* (pp. 1–6). IEEE.

Chen, Z., Renda, F., Le Gall, A., Mocellin, L., Bernabei, M., Dangel, T., Ciuti, G., Cianchetti, M., & Stefanini, C. (2023). Data-driven methods applied to soft robot modeling and control: A review. *IEEE Transactions on Automation Science and Engineering*, *20*(3), 1433–1454.

Díaz Rodríguez, N., Ser, J., Coeckelbergh, M., López de Prado, M., Herrera-Viedma, E., & Herrera, F. (2023). Connecting the dots in trustworthy artificial intelligence: From AI principles, ethics, and key requirements to responsible AI systems and regulation. *Information Fusion*, *99*, 101869.

Farkh, R., & Aljaloud, K. A. (2022). Vision navigation based PID control for line tracking robot. *Intelligent Automation & Soft Computing*, *34*(3), 1307–1324.

Foughali, M., & Zuepke, A. (2022). Formal verification of real-time autonomous robots: An interdisciplinary approach. *Frontiers in Robotics and AI*, *9*, 847894.

Ghzouli, R., Berger, T., Johnsen, E. B., Wąsowski, A., & Dragule, S. (2022). Behavior trees and state machines in robotics applications. *IEEE Transactions on Software Engineering*, *48*(12), 5051–5074.

González-Santamarta, M., Rodríguez-Lera, F. J., Fernández-Llamas, C., & Matellán-Olivera, V. (2024). A hybrid cognitive architecture to generate, control, plan, and monitor behaviors for interactive autonomous robots. *International Journal of Social Robotics*. *17*(12), 3135–3151.

Hegde, C., Su, J., Tan, J. M. R., He, K., Chen, X., & Magdassi, S. (2023). Sensing in soft robotics. *ACS Nano*, *17*(15), 14159–14188.

Hughes, N., Chang, Y., Hu, S., Talak, R., Abdulhai, R., Strader, J., & Carlone, L. (2024). Foundations of spatial perception for robotics: Hierarchical representations and real-time systems. *The International Journal of Robotics Research*, *43*(10), 1457–1505.

Ji, T., Sivakumar, A. N., Chowdhary, G. V., & Driggs-Campbell, K. (2022). Proactive anomaly detection for robot navigation with multi-sensor fusion. *IEEE Robotics and Automation Letters*, *7*(3), 7299–7306.

Karle, P., Fent, F., Huch, S., Sauerbeck, F., & Lienkamp, M. (2023). Multi-modal sensor fusion and object tracking for autonomous racing. *IEEE Transactions on Intelligent Vehicles*, *8*(3), 2163–2177.

Kawaharazuka, K., Oh, J., Yamada, J., Posner, I., & Zhu, Y. (2025). Vision-language-action models for robotics: A review towards real-world applications. *IEEE Access*, *13*, 150233–150268.

Kondratenko, Y., Atamanyuk, I., Sidenko, I., Kondratenko, G., & Sichevskyi, S. (2022). Machine learning techniques for increasing efficiency of the robot's sensor and control information processing. *Sensors*, *22*(3), 1130.

Leal, H. M., Barbosa, R. S., & Jesus, I. S. (2025). Control of a mobile line-following robot using neural networks. *Algorithms*, *18*(1), 23.

Macenski, S., Foote, T., Gerkey, B. P., Lalancette, C., & Woodall, W. (2022). Robot operating system 2: Design, architecture, and uses in the wild. *Science Robotics*, *7*(66), eabm6074.

Maken, F. A., Muthu, S., Nguyen, C., Sun, C., Tong, J., Wang, S., Tsuchida, R., Howard, D., Dunstall, S., & Petersson, L. (2025). Improving 3D reconstruction through RGB-D sensor noise modeling. *Sensors*, *25*(3), 1011.

Meli, D., Nakawala, H., & Fiorini, P. (2023). Logic programming for deliberative robotic task planning. *Artificial Intelligence Review*, *56*(7), 6551–6598.

Milojevic, D., Zardini, G., Elser, M., Censi, A., & Frazzoli, E. (2025). CODEI: Resource-efficient task-driven codesign of perception and decision making for mobile robots applied to autonomous vehicles. *IEEE Transactions on Robotics*, *41*(2), 987–1005.

Mittal, M., Yu, C., Yu, Q., Liu, J. J., Rudin, N., Hoeller, D., Yuan, J., Poorsarvi Tehrani, P., Singh, R., Guo, Y., Mazhar, H., Mandlekar, A., Babich, B., State, G., Hutter, M., & Garg, A. (2023). Orbit: A unified simulation framework for interactive robot learning environments. *IEEE Robotics and Automation Letters*, *8*(2), Toy–T20.

Patton, N. T., Rahmani, K., Missula, M., Biswas, J., & Dillig, I. (2023). Programming-by-demonstration for long-horizon robot tasks. *Proceedings of the ACM on Programming Languages*, *7*(PLDI), 1523–1550.

Rizwan, M., Reichenbach, C., Caldas, R., Mayr, M., & Krüger, V. (2025). EzSkiROS: Enhancing robot skill composition with embedded DSL for early error detection. *Frontiers in Robotics and AI*, *11*, 1287654.

Somesh, D. H., & Aruna, R. (2025). Line follower robot. *International Journal of Advanced Research in Science, Communication and Technology*, *6*(2), 98–105.

Vemprala, S. H., Bonatti, R., Bucker, A., & Kapoor, A. (2023). ChatGPT for robotics: Design principles and model abilities. *IEEE Access*, *11*, 49645–49663.

5 Software Engineering for Robotics

5.1 INTRODUCTION

Programming intelligent robotic systems is only one piece of the puzzle. Software engineering will help your team build robotic systems that can scale, are easy to maintain, and allow your team to collaborate effectively (Caldas et al., 2024). Robotic systems tend to become more complicated as more features are added, consisting of many different subsystems that need to work together reliably (Yang & Zhang, 2023). Bad software architecture makes it hard to keep systems running, reason about how to update a large system, and troubleshoot software bugs (Lumpp et al., 2023).

Robotics software can involve sensing, make decisions, and actuate one or more pieces of hardware. These components are typically run simultaneously and need to communicate with each other. Debugging your code becomes more difficult, and the software architecture is more complex than a single program running on a computer (Macenski et al., 2022). For these reasons, it is important to build robot software with a structured design (Podlubne et al., 2024).

Modularity refers to the idea of breaking something complex into smaller parts. Doing this with your software makes it easier to test, update, and reuse parts of your system (Xu & Li, 2022). Robotics applications are usually never done by a single engineer. Software engineering encompasses techniques that can help teams collaborate on code, such as using version control systems, writing documentation, and communicating about your code (Caldas et al., 2024; El Hafi et al., 2022). Robot Operating System (ROS) is a robotics framework that was designed around many of these ideas, like allowing you to write modular and distributed systems that can communicate with one another (Macenski et al., 2022).

5.2 FOUNDATIONS OF SOFTWARE ENGINEERING IN ROBOTICS

Software engineering best practices define how programs should be built so that they become reliable, scalable, and last over time (Macenski et al., 2022). Robotics programs generally tend to become complex systems consisting of various subsystems instead of being single programs. Some foundational rules should be followed when architecting software systems so that with growth, complexity does not become unmanageable (Ye et al., 2023).

Programs should be built using modular designs, where the entire pipeline is split into separate sections or nodes, with each responsible for one aspect of the job. Common modular design examples are modules for sensing, perception/navigation, and acting (Macenski et al., 2022; Xu & Li, 2022). Breaking down a project into

DOI: 10.1201/9781003758501-7

modules allows for independent development, debugging, and code replacement. Modules can also be designed to be reused between projects (Lumpp et al., 2023).

Code should be easy to change as the robot program is being improved over time. Well-written code that is easy to read and understand will decrease development time. Code should use easy-to-understand naming, be neatly organised, and be properly commented to allow for faster comprehension of what the code is intended to do (Yang & Zhang, 2023).

Programs should be written with reuse in mind so that parts of them can be used down the line for other projects (Yang & Zhang, 2023).

Code should always be tested to verify that the output of a node or an entire pipeline is acting as expected. These tests should be run repeatedly to catch any regressions (Caldas et al., 2024). Following software engineering best practices will allow you to create rock solid robotics software (Valner et al., 2022).

5.3 STRUCTURING ROBOTIC SYSTEMS WITH ROS

Creating a multi-component robot demands an underlying structure that allows communication, coordination, and action-taking capabilities to be shared between components. ROS is an open-source architecture with a collection of tools and libraries designed to develop robotics systems in a modular and distributed fashion (Macenski et al., 2022). ROS is a middleware program and not an actual operating system (Lienen et al., 2023).

Nodes are processes written in ROS that perform a task. Actions can be split between nodes depending on the design of the robot. For instance, sensor readings can be managed by one node, processing by another, and actuator control by another (Bonci et al., 2023). Nodes allow for separation of functionalities into independent processes that can work together to dictate global robot behaviours (Lienen et al., 2023).

ROS facilitates communications between nodes by using topics and messages. Topics are named buses over which nodes exchange messages using a publish–subscribe data transmission model (Macenski et al., 2023). ROS features request–response functionality through services. Nodes can directly request information from other nodes through a service when necessary.

Figure 5.1 depicts three example nodes, three topics, and three messages. Sensor nodes provide information, process nodes receive information (publish subscribe) from sensors and provide processed information, and control nodes use processed

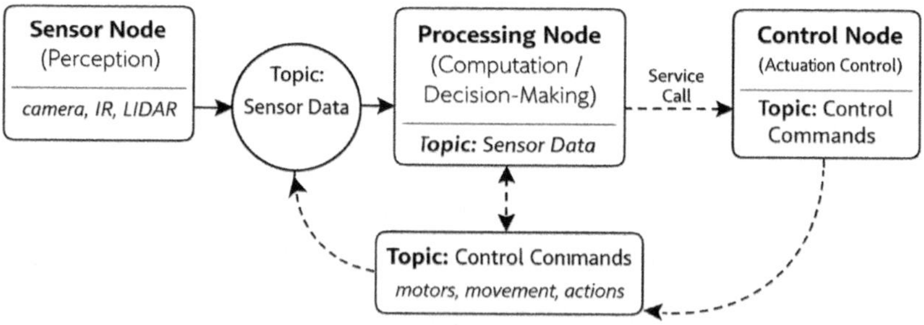

FIGURE 5.1 ROS-Based control system architecture.

information to send out commands. Robots can be segmented into these nodes to create more complex systems.

Nodes, topics, and messages allow systems to scale as nodes can be added and created without changing current nodes. Updates can also be made to support different hardware, allowing flexibility in development. ROS also allows for multiple people to work on the same robot. ROS also allows us to break up the complexity of robots into smaller segments that are easily changed, updated, and understood.

5.4 CASE STUDY: MODULAR ROS SYSTEM DEVELOPMENT USING AGILE

Designing large robotic systems often necessitates teamwork and sound software engineering principles such as modular design and continuous integration (El Hafi et al., 2022; Macenski et al., 2022). In this case study, we demonstrate the design and implementation of such a robot controller system through team projects. Groups of students work together to design individual components and incrementally implement them in ROS, while employing agile software development practices to manage the project (Edison et al., 2022; Vieira et al., 2025).

The robot controller software is split into several modules that can each run as independent processes, organised as ROS nodes (Lumpp et al., 2023; Macenski et al., 2022). For instance, one node could provide perception capabilities to interpret sensor data, while another node plans paths for the robot to take and control its motors (Caldas et al., 2024). Nodes can communicate with each other over topics by publishing and subscribing to messages, facilitating loose coupling between components (Caldas et al., 2024; Macenski et al., 2022). For example, a perception node may read in raw sensor data and publish processed information to be used by a controller node to generate commands to send to the robot's motors (Vieira et al., 2025). Splitting the controller software into loosely coupled nodes allows for separation of concerns and easier maintenance by multiple people at the same time.

The project is structured around agile development practices, with tasks grouped into time-boxed sprints (Edison et al., 2022). Student teams break down desired robot behaviours into user stories, then estimate and prioritise work to be completed as items in a product backlog (Lee & Chen, 2023). Team members hold each other accountable by meeting frequently to demonstrate the work they have completed and integrating their code together to test the system (Strode et al., 2022).

Integration and frequent testing help to ensure the system is performing as expected and help catch issues early on to minimise risk (Caldas et al., 2024; Donca et al., 2022). Teams can easily iterate on parts of the system they are responsible for, such as using a higher resolution camera, without needing to redesign the entire system architecture (El Hafi et al., 2022). This case study has illustrated how software engineering principles allow groups of people to collaborate and build sophisticated robots.

5.5 SOFTWARE DEVELOPMENT LIFE CYCLE (SDLC) IN ROBOTICS

SDLC stands for software development life cycle (Donca et al., 2022). This development method defines stages of a project that help software developers design systems. Robots are complex machines with software and hardware parts. Robotics engineers can use the SDLC to build error-free reliable systems with reusable code (Xu & Li, 2022). The modular design will allow you to scale your project in the future.

Requirements analysis is defining the purpose and limitations of your system. You should know what your robot will do, what sensors and actuators will be needed and under what conditions the robot will work (Xu & Li, 2022). With these requirements gathered, you can ensure your project will have consistency across all stages of development (Araujo et al., 2022).

System design is the stage in which your system's architecture is chosen (Podlubne et al., 2024). For robotics projects, you can decide what packages will act as ROS nodes and what communication interfaces will be used (Macenski et al., 2022). Besides that, how each subsystem will interact on a system level should also be defined (Lumpp et al., 2023). Designing good system architecture is vital to producing quality results.

Implementation is the stage in which you will write code (Podlubne et al., 2024). Software for each node is created along with any integration programs (Donca et al., 2022). Bugs are fixed so the system operates as intended. In robotics, you should also make sure your code interfaces with your physical system correctly.

Testing helps you make sure your system operates as expected (Araujo et al., 2022). Test each individual piece of your system to ensure every link in the chain functions (Caldas et al., 2024). Finally, you should test your system in its working environment. Run through your software with your robot to check that everything works safely.

Deployment and maintenance are the stages of releasing your system and improving it (Valner et al., 2022). Update and fix bugs in your system as you see needed. You can continually improve your system's performance during this stage (Macenski et al., 2022).

5.6 OBJECT-ORIENTED PROGRAMMING IN ROBOTICS

Object-oriented programming (OOP) is a programming paradigm that models software as objects that contain data and behaviour (Vera Vera & Vera Vera, 2024). In robotics, OOP can be used to create modular and easily understandable systems. The building blocks of OOP are classes and objects (Liu, 2025). Classes are like blueprints for objects and define their properties and behaviour. Objects are instances of classes that can interact with each other. For example, in robotics, a class could represent a robot arm, while an object could represent a specific arm.

Encapsulation is the practice of grouping data and functions that manipulate that data into a single unit called an object. Classes provide a way to encapsulate data and methods into reusable components (Jimmy, 2025). With encapsulation, programmers can hide implementation details and only expose methods that are necessary to use the object. In robotics, this allows for the reuse of components without having to know how they are implemented internally.

Inheritance allows programmers to create new classes that are based on existing classes. Classes can inherit attributes and behaviours from other classes and add or override as needed. This can be useful in robotics when creating specialised components that share some common functionality (Liu, 2025). For example, if there are multiple types of sensors on a robot, a sensor class could be created with basic functions. Then, each specific sensor type can inherit from the sensor class and add their own unique functionality.

Abstraction is the process of reducing complexity by focusing on the essential qualities of something while ignoring irrelevant details (Lavand & Pawar, 2025). In OOP, abstraction can be used to represent real-world objects or components of a robotic system at a higher level. Abstraction allows programmers to create classes that represent these components without having to worry about the implementation details.

Since robots are made up of multiple components that work together, using OOP allows these components to be programmed to interact with each other using defined interfaces. Robots typically have a modular design which allows OOP to fit nicely. ROS even uses similar concepts of encapsulation in their framework, which allows for robots to be scalable (Yang & Zhang, 2023).

5.7 VERSION CONTROL AND COLLABORATIVE DEVELOPMENT

Version control refers to keeping track of changes made to your code (Ghodke & Chavan, 2024). It helps you see what has changed over time, who changed it, and why. Version control is necessary when multiple engineers work on the same robotic system. Collaborating on software projects requires a mechanism for controlling when and how new contributions get added to the codebase (El Hafi et al., 2022).

Git is a version control system that helps you keep track of who made changes to what files and when (Ghodke & Chavan, 2024). Every change set to your project is stored as a commit object, which allows you to go back and look at old versions of your project files. As you make edits, you can also see who last modified each line if something goes wrong.

Branching in Git lets you copy a piece of code so that you can work on something else without affecting the main project. For instance, one developer can create a branch to work on their navigation module while another teammate works on the sensor interface code. Branches can be merged to add changes from one line of development to another (Mamidi, 2025).

Collaboration occurs when two or more engineers make changes to the same section of code (Shen et al., 2022). Merge conflicts occur when Git can't automatically reconcile two changes made to a file (Vale et al., 2023). When collaborators edit the same block of code, a conversation needs to take place to sort out these conflicts (Ribeiro et al., 2022). When merging their branches, teammates will have to manually resolve these issues.

Engineering teams can collaborate by following processes around how new code is contributed to their projects. Working together may involve pulling requests, reviewing each other's code, and testing new features before merging them to main. Teammates who adopt software engineering best practices like this tend to have fewer bugs in their codebase (Herrera González & Fraternali, 2023).

Version control is critical for robotics because software often interfaces directly with hardware (El Hafi et al., 2022). You must be able to track how code changes may impact the behaviour of your project (Coronado et al., 2024). Tools like Git allow robotics teams to coordinate their efforts and control how their system evolves over time (Ghodke & Chavan, 2024).

5.8 AGILE DEVELOPMENT FOR ROBOTICS

Agile development refers to iterative software development practices with short cycles (aka "sprints") and puts a heavy focus on communication among team members. Agile development lends itself quite nicely to robotics application development because robotic systems frequently have requirements that change or are realised during testing and system deployment (Edison et al., 2022; El Hafi et al., 2022).

User stories and tasks are organised by development sprint. Each sprint is a period during which an agile development team attempts to complete a subset of tasks on the product backlog, which is a prioritised list of all the work that needs to be done (Lieberum et al., 2022). The product should be in a stable state at the end of each sprint. When developing robots, this means that the robot should be operable with the newly implemented features from that sprint (Lee & Chen, 2023). Tasks are created as user stories, which typically represent some new behaviour of the system (Lee & Chen, 2023). These behaviours can include processing sensor data, motion control, or even new robot actions (Caldas et al., 2024; El Hafi et al., 2022).

Product backlogs are living documents that will grow and change as development continues. The backlog allows the team to continually reprioritise future tasks as the project evolves. At the beginning of each sprint, the team decides which stories they will attempt to complete by the end of that sprint (Lieberum et al., 2022). Since communication is key in Agile development, frequent meetings like daily updates can help teams stay aligned (Behutiye et al., 2022).

Code is integrated and tested frequently during agile software development. Testing is performed in simulation, test beds, or even the real world, which allows teams to discover errors quickly and limit overall project risk (Donca et al., 2022).

Agile development's biggest strength in robotics is its flexibility. As new requirements are discovered, the backlog can be updated, and new changes can be implemented at the start of their respective sprint (Lee & Chen, 2023). Overall, Agile methodology allows for incremental development, collaboration, and management of robotics systems (Edison et al., 2022; El Hafi et al., 2022).

5.9 PRACTICAL INSIGHT: MANAGING A ROBOTICS PROJECT

Building and managing a robotic project involves software development practices such as team communication and Agile software development principles (El Hafi et al., 2022). First, as a team, it is essential to gather around a project board and decide what the robotic system should do (Dong et al., 2024). Common robotic projects include mobility, sensing, and grasping. These ideas are further refined into working software requirements and drafted into tasks (Behutiye et al., 2022).

These smaller tasks are formulated as user stories that describe how the system should behave at a higher level (Lee & Chen, 2023). Tasks are then grouped into what is called a "sprint." During sprint planning, the team decides what changes will be made to the system during development (Edison et al., 2022). For example, the team could decide to implement sensor reading during one sprint and movement control during another.

Tasks are then completed by team members, usually through the development of code (El Hafi et al., 2022). Version management platforms such as Git are helpful when many collaborators edit code frequently (Lee & Chen, 2023). Communication with team members during this process helps the team avoid integrating incompatible components. The new components are also incrementally integrated into the system to verify their functionality and evaluate overall system behaviour.

System test cases are designed and run to ensure the system meets requirements (Araujo et al., 2022). Testing occurs at the module level first and progresses to more significant integrations as the system begins to take shape. As more functionality is added to the system, testing focuses less on module functionality and more on overall system behaviour (El Hafi et al., 2022).

As issues are discovered or encountered by the team, it is important to manage these risks by adapting the project design and rollout schedule (Binboga & Gumussoy, 2022). Agile software development allows a team to flex their planning to the project's needs. If an aspect of the project is taking too long or is not functioning as desired, then the team can adapt their plan to address these shortcomings (Dong et al., 2024). Meeting at the end of each sprint to discuss what occurred is known as a retrospective.

The above example provides a real-world instance of how software engineering is used to build projects (El Hafi et al., 2022). Software engineering combines smaller procedures to allow large teams to create complex systems in a manageable way (Edison et al., 2022).

5.10 CHALLENGES IN ROBOTICS SOFTWARE ENGINEERING

Software development for robotics systems is more challenging than traditional software development for several reasons.

Robotics systems operate in the real world and need to integrate perception, decision-making, and actuation (Caldas et al., 2024). Robotics software is often complex and modular, with real-time constraints. Therefore, robotics systems need to be engineered robustly (Gambo et al., 2025).

System integration challenges can occur as robotics systems combine different modules for perception, planning, navigation, control, etc. Middleware like ROS helps abstract hardware but guaranteeing compatibility between software modules can be challenging. Sometimes individual components work separately but may cause unforeseen bugs when linked together (Gambo et al., 2025).

Robotics software needs to be coupled with hardware limitations like sensing/measurement accuracy, actuation limits, computation speeds, and latency. Code that runs fine in simulation may not work or behave as expected when run by physical robots (Caldas et al., 2024).

Real-time constraints are also important to consider. Robots need to perceive their environment, plan, and take actions quickly enough to be effective. Computational delays can cause robots to run into things or fall over (Jin et al., 2025).

Robotics systems are usually large and require teamwork to build. Splitting up a project into smaller tasks can pose challenges with team members' code being incompatible or working on the same things without realising it (El Hafi et al., 2022).

5.11 ETHICAL AND REAL-WORLD CONSIDERATIONS

Robotics systems are often developed through collaborative software engineering efforts. Shared responsibility has consequences that need to be considered when building software that performs real-life functions. Code quality, reliability, and maintainability must be considered when writing software (Araujo et al., 2022). Robot behaviours are translated through software, so sub-par coding can cause unsafe robot behaviours. When creating software for robots that work with or around humans, or can operate autonomously, great care should be taken to ensure responsible and professional software engineering practices are being used (Araujo et al., 2022).

Robotics engineers frequently use code found online to help speed up the development process. Often, these code bases are open source and lack proper documentation. Using code from tutorials, forums, and mainly open-source frameworks like ROS can lead to tough debugging sessions if the code is not what it seems (El Hafi et al., 2022; Macenski et al., 2022). Some problems can be hard to detect when writing software, so it is important to be aware of what your code is using and why you chose to use it (Yumbla et al., 2023).

Operating in unknown environments can lead to unpredictable situations. If something were to go wrong in the robot, how would the system handle it? Would it stop moving altogether or try to recover on its own? Safety should always be the priority (Weidemann et al., 2023). Programming with best practices and design patterns can help prevent unwanted software behaviours. Considering these ideas when writing software will help you become a better software engineer and roboticist. It is vital not to simply write working code, but responsible code (Othman & Yang, 2023).

5.12 CONCLUSION

This chapter introduced software engineering concepts relevant to building practical robotic systems. Robot applications increasingly require disciplined software engineering methods in order to scale complex systems. Robot applications that were previously programmed as simple scripts now require the rigour of engineering to scale to real-world applications. Throughout this chapter, we have covered concepts related to modularity, reuse, and maintainability of code. Frameworks like ROS allow for distributed systems composed of many subsystems that communicate with one another. ROS also allows us to develop, test, and improve each subsystem in isolation.

Software engineering practices like version control and Agile allow teams to work together to make changes to a system. Version control allows teams to stay coordinated when changing code. Agile software development techniques allow systems to be continuously improved as requirements change or become clearer. SDLC provides a template for taking a system from conceptualisation all the way to deployment and maintenance.

5.13 DISCUSSION QUESTIONS

1. What is software engineering and why do we need it when building complex robots?
2. How does modular design enable better scalability in robotics software?
3. Describe how ROS enables distributed system design.
4. What is version control and how can we use it when working together?
5. How does Agile methodology benefit us when working on robotics projects?

5.14 EXERCISES

Exercise 1: Sprint Planning

Plan out what a sprint would look like for a smaller robotics project. For example, a project could be building a line-following robot. Come up with at least three user stories to include in a one-week sprint.

Exercise 2: Version Control Roleplay

Come up with a scenario to show how two developers would work on their own branch to implement separate features of code. Show how they would merge their code together and what to do if there are conflicts.

Exercise 3: ROS Roleplay

Draw out an example ROS system. Come up with at least three nodes that would work together (i.e., sensor node, processing node, control node). Describe how the nodes will communicate using ROS topics.

Exercise 4: Reflecting on Team Development

Why is it difficult to work as a team to develop a robotic system? Come up with ideas on how to collaborate effectively and minimise errors.

REFERENCES

Araujo, H. L. S., Mousavi, M., & Varshosaz, M. (2022). Testing, validation, and verification of robotic and autonomous systems: A systematic review. *ACM Transactions on Software Engineering and Methodology*, *31*(4), 1–64.

Behutiye, W., Tripathi, N., & Isomursu, M. (2022). Adopting scrum in hybrid settings, in a university course project: Reflections and recommendations. *IEEE Access*, *10*, 120186–120203.

Binboga, B., & Gumussoy, C. A. (2022). Factors affecting agile software project success. *IEEE Access*, *10*, 63784–63802.

Bonci, A., Gaudeni, F., Giannini, M. C., & Longhi, S. (2023). Robot Operating System 2 (ROS2)-based frameworks for increasing robot autonomy: A survey. *Applied Sciences*, *13*(23), 12828.

Caldas, R. D., Pinera García, J. A., Schiopu, M., Pelliccione, P., Rodrigues, G., & Berger, T. (2024). Runtime verification and field-based testing for ROS-based robotic systems. *IEEE Transactions on Software Engineering*, *50*(4), 1172–1194.

Coronado, E., Ueshiba, T., & Ramirez-Alpizar, I. G. (2024). A path to Industry 5.0 digital twins for human–robot collaboration by bridging NEP+ and ROS. *Robotics*, *13*(1), 9.

Donca, I., Stan, O., Misaros, M., Goţa, D., & Miclea, L. (2022). Method for continuous integration and deployment using a pipeline generator for agile software projects. *Sensors*, *22*(11), 4133.

Dong, H., Dacre, N., Baxter, D., & Ceylan, S. (2024). What is agile project management? Developing a new definition following a systematic literature review. *Project Management Journal*, *55*(3), 213–232.

Edison, H., Wang, X., & Conboy, K. (2022). Comparing methods for large-scale agile software development: A systematic literature review. *IEEE Transactions on Software Engineering*, *48*(12), 4886–4917.

El Hafi, L., Garcia Ricardez, G. A., von Drigalski, F., Inoue, Y., Yamamoto, M., & Yamamoto, T. (2022). Software development environment for collaborative research workflow in robotic system integration. *Advanced Robotics*, *36*(11–12), 564–582.

Gambo, M. L., Danasabe, A., Almadani, B., Aliyu, F., Aliyu, A., & Al-Nahari, E. (2025). A systematic literature review of DDS middleware in robotic systems. *Robotics*, *14*(3), 67.

Ghodke, G. M., & Chavan, T. (2024). An overview of Git. *International Journal of Scientific Research in Modern Science and Technology*, *5*(3), 10–18.

Herrera González, S. L., & Fraternali, P. (2023). Almost Rerere: Learning to resolve conflicts in distributed projects. *IEEE Transactions on Software Engineering*, *49*(7), 3843–3868.

Jimmy, J. (2025). Design and implementation of a file encryption system based on object-oriented programming using C++. *Journal of Computer Networks, Architecture and High-Performance Computing*, *7*(4), 45–55.

Jin, H., Wu, J., Yuan, W., Liu, F., & Cui, Y. (2025). Co-design of sensing, communications, and control for low-altitude wireless networks. *IEEE Transactions on Mobile Computing*, *24*(7), 4201–4217.

Lavand, M., & Pawar, R. (2025). A study on abstraction in object-oriented programming with Java and its implementation in developing programs. *International Journal on Science and Technology*, *5*(2), 10–16.

Lee, W.-T., & Chen, C.-H. (2023). Agile software development and reuse approach with Scrum and software product line engineering. *Electronics*, *12*(15), 3220.

Lieberum, T., Schiffels, S., & Kolisch, R. (2022). Should we all work in sprints? How agile project management improves performance. *Manufacturing & Service Operations Management*, *24*(3), 1254–1273.

Lienen, C., Nowosad, A. P., & Platzner, M. (2023). Mapping and optimizing communication in ROS 2-based applications on configurable system-on-chip platforms. Proceedings of the 2023 9th International Conference on Robotics and Artificial Intelligence.

Liu, D. (2025). Design and implementation of an intelligent robot control system based on Java inheritance and polymorphism characteristics. *Education Reform and Development*, *7*(3), 101–115.

Lumpp, F., Panato, M., Bombieri, N., & Fummi, F. (2023). A design flow based on Docker and Kubernetes for ROS-based robotic software applications. *ACM Transactions on Embedded Computing Systems*, *22*(3), 1–27.

Macenski, S., Foote, T., Gerkey, B. P., Lalancette, C., & Woodall, W. (2022). Robot Operating System 2: Design, architecture, and uses in the wild. *Science Robotics*, *7*(66), eabm6074.

Macenski, S., Soragna, A., Carroll, M., & Ge, Z. (2023). Impact of ROS 2 node composition in robotic systems. *IEEE Robotics and Automation Letters*, *8*(7), 3900–3907.

Mamidi, S. (2025). The Git branching strategy: A comprehensive guide. *International Journal for Multidisciplinary Research*, *7*(11), 45–60.

Othman, U., & Yang, E. (2023). Human–robot collaborations in smart manufacturing environments: Review and outlook. *Sensors*, *23*(11), 5207.

Podlubne, A., Mey, J., Andreou, A., Pertuz, S., Aßmann, U., & Göhringer, D. (2024). Model-based generation of hardware/software architectures with hybrid schedulers for robotics systems. *IEEE Transactions on Computers*, *73*(7), 1848–1862.

Ribeiro, B. B., Costa, C., & Santos, R. (2022). Understanding and analyzing factors that affect merge conflicts from the perspective of software developers. *Journal of Software Engineering Research and Development*, *10*, 9.

Shen, B., Gulzar, M. A., He, F., & Meng, N. (2022). A characterization study of merge conflicts in Java projects. *ACM Transactions on Software Engineering and Methodology*, *31*(3), 1–41.

Strode, D., Dingsøyr, T., & Lindsjørn, Y. (2022). A teamwork effectiveness model for agile software development. *Empirical Software Engineering*, *27*(4), 94.

Vale, G., Costa, H. A. X., & Apel, S. (2023). Predicting merge conflicts considering social and technical assets. *Empirical Software Engineering*, *28*(6), 142.

Valner, R., Vunder, V., Aabloo, A., Pryor, M., & Kruusamäe, K. (2022). TeMoto: A software framework for adaptive and dependable robotic autonomy with dynamic resource management. *IEEE Access*, *10*, 103042–103060.

Vera Vera, J. B., & Vera Vera, J. R. (2024). The role of object-oriented programming in sustainable and scalable software development. *Minerva*, *3*(1), 45–60.

Vieira, D., Oliveira, M., Arrais, R., & Melo, P. (2025). Application of cloud simulation techniques for robotic software validation. *Sensors*, *25*(5), 2150.

Weidemann, C., Mandischer, N., van Kerkom, F., Corves, B., Hüsing, M., Kraus, T., & Garus, C. (2023). Literature review on recent trends and perspectives of collaborative robotics in Work 4.0. *Robotics*, *12*(3), 78.

Xu, R., & Li, C. (2022). A modular agricultural robotic system (MARS) for precision farming: Concept and implementation. *Journal of Field Robotics*, *39*(4), 551–571.

Yang, S., & Zhang, Q. (2023). Towards efficient robotic software development by reusing behavior tree structures for task planning paradigms. *Complex Systems Modeling and Simulation*, *3*(2), 129–159.

Ye, Y., Nie, Z., Liu, X., Xie, F., Li, Z., & Li, P. (2023). ROS2 real-time performance optimization and evaluation. *Chinese Journal of Mechanical Engineering*, *36*(1), 104.

Yumbla, F., Fajardo-Pruna, M., Piguave, A., Ronquillo, D., Ortiz, R., Choi, J., Diaz, G., Pañeda, X. G., & Moon, H. (2023). An open-source multi-robot framework system for collaborative environments based on ROS2. *IEEE Access*, *11*, 141088–141104.

6 Securing Robotic Systems

6.1 INTRODUCTION

Robotics can be defined as mechanisms that can act in environments such as homes, offices, or factories (Botta et al., 2023). Robots range from humanoid robots to industrial machinery. Robots are getting smarter every day with more processing and sensing capabilities coupled with more connected communication systems. This combination has changed robots from closed systems to cyber-physical systems susceptible to cybersecurity attacks (Tanimu & Abada, 2024).

Robot systems include software, hardware, and connecting networks which create multiple avenues for attack. Attacks can range from access to sensitive information passing through a robot to complete takeover of the robot's behaviour (Bhardwaj et al., 2025; Shaik et al., 2025). Robots used for interacting with humans can share private information if attacked or could cause harm to their environment or users (Jiang et al., 2024). While these vulnerabilities exist in cybersecurity, they have been known and patched for a while. The difference in applying these concepts to robotics is that they can cause physical-world behaviour (Haskard & Herath, 2025; Xing & Shen, 2024).

Security of robotic systems should be considered from the ground up in their design to prevent data breaches, information leakages, and safety concerns (Haskard & Herath, 2025). Securing robots has become crucial as they are used to interact with humans. If security vulnerabilities are found, attackers can use a robot's behaviour to harm others (Callander et al., 2024; Shaik et al., 2025; Wu et al., 2023).

This chapter will cover cybersecurity concepts and how they map to robotics. We will discuss these concepts and how to build safe, secure, and ethical systems (Tanimu & Abada, 2024).

6.2 FOUNDATIONS OF SECURITY IN ROBOTICS

Security, by definition, is ensuring data and systems are safe from threats that try to breach them without permission (Bhardwaj et al., 2025; Botta et al., 2023). Since robots are cyber-physical systems that collect data and physically interact with the world based on that data, we must take both confidentiality and integrity into consideration (Yu et al., 2023). One common framework used when discussing security is called the Confidentiality, Integrity, and Availability triad or CIA triad (Jeffrey et al., 2023).

Confidentiality is about making sure that information is not seen by unauthorised individuals or systems. Many robotic systems process information that could be sensitive, such as private user data, sensor information, or internal system processes (Botta et al., 2023). Suppose someone were to see this data, like what objects are in a user's home from their home robot. In that case, it could cause privacy issues for that user. Confidentiality can be assured using methods like encryption or authorisation (Jiang et al., 2024).

DOI: 10.1201/9781003758501-8

Integrity is about making sure that our information and systems are accurate and have not been tampered with. Robotic systems heavily rely on integrity since if our data is wrong, our robot's actions may be unsafe. For example, if someone were to mess with the sensors on a robot, they can provide wrong information to the robot. This can cause the robot to think that there is something in its path when there is none (Xing & Shen, 2024). Integrity can be maintained by using validation, authentication, and encryption (Jiang et al., 2024; Yu et al., 2023).

Availability is about making sure that our systems and data can be accessed when we need to. Robotic systems need to be available to operate; this is especially true for real-time systems (Jeffrey et al., 2023). If someone were to take down our robot's servers, we would not be able to use our robot. These types of attacks can be dangerous since we won't be able to use the system when we need to (Zhang et al., 2024).

The CIA triad is a great tool to use when analysing our system's security. It allows us to break down security into three important aspects that we can focus on when creating our systems (Bhardwaj et al., 2022). Ensuring confidentiality, integrity, and availability will allow our system to withstand many types of attacks and allow it to operate as intended.

6.3 MODELLING THREATS IN ROBOTIC SYSTEMS

Threat modelling is a technique used to understand potential threats to a system, including how they may exploit vulnerabilities and what damage they can cause (Bhardwaj et al., 2022; Yu et al., 2023). Threat modelling also helps to identify countermeasures that can be put in place to mitigate these threats (Sheikh & Singh, 2022). In robotics, threat modelling helps identify and address security risks associated with robotic systems.

Robotic systems typically include hardware, software, and communication networks (Botta et al., 2023). All of these can represent attack surfaces that can be exploited by malicious actors. Threat modelling typically starts by identifying assets in the system (Phillips et al., 2022). Assets represent something that we wish to protect. In robotic systems, assets can include sensor information, user data, commands sent to the robot, and communications between robots or with controllers.

After assets have been identified, potential threats to those assets must be identified (Meland et al., 2022). Threats can come from external sources like hackers or inside attackers who have access to the robot or network. Some examples of threats are reading information from the controller memory that could allow for remote code execution commands or extracting sensor information that could allow spoofing of those sensors.

Next, vulnerabilities must be identified (Amro et al., 2022). Vulnerabilities represent weaknesses in a system that could be exploited by threats to gain access to assets. This could take many forms, such as a lack of encryption when data is sent over communication networks or a lack of authentication when commands are issued. The last step is determining how an attacker can exploit system vulnerabilities and impact assets.

Impact represents the harm that can be caused if a threat actor successfully exploits a vulnerability (Yu et al., 2023). Typical impacts include loss of confidentiality,

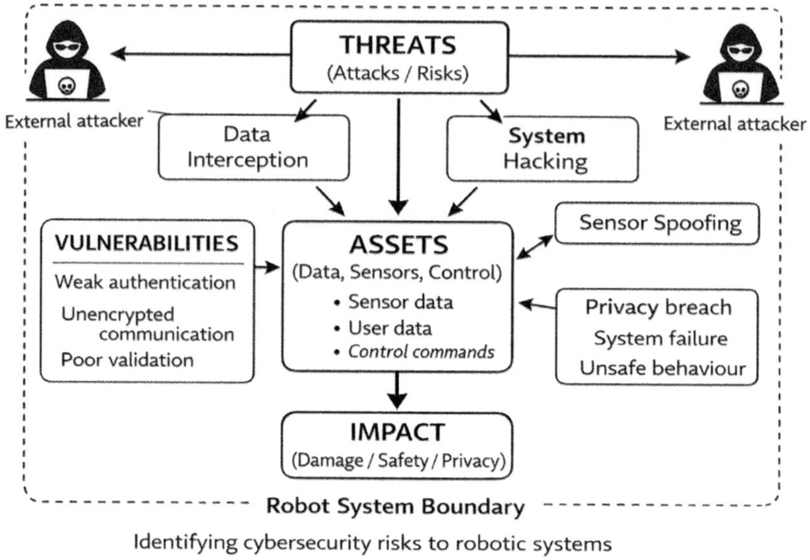

FIGURE 6.1 Threat model for a robotic system.

integrity, and availability of data. Robot failures can cause more serious impacts than just loss of information. If a robot gets hacked, it can cause it to operate incorrectly and hurt people.

Figure 6.1 shows how assets, threats, vulnerabilities, and impacts can relate to one another within robotics. Mapping these details allows roboticists to address various failure points with security implementations. Threat modelling allows security to be designed ahead of time, ensuring robotic applications are thoroughly thought out prior to release.

6.4 CASE STUDY: XIAOMI ROBOT VACUUM DATA INCIDENT

Robot systems have been flooding our homes, and privacy advocates have raised concerns regarding how such systems manage users' data, keeping them safe from malicious attacks. One case relates to the security concerns raised by Xiaomi robot vacuums (Ulsmåg et al., 2024).

Robot vacuums like Roomba use proximity sensors and programming mapping algorithms to avoid collisions with walls/furniture and navigate around the rooms of their owners. While they typically collect data about room layouts and possible locations of their users to operate more effectively and avoid obstacles, such data might sometimes be considered sensitive (Song et al., 2024; Ulsmåg et al., 2024). This type of data leak may allow outsiders to collect information about people's homes and possible daily movement routines inside their property (Grabler & Koeszegi, 2024).

Reports show that Xiaomi robot vacuum leaked such information to external servers (Ulsmåg et al., 2024). Since leaking such sensitive data compromises data confidentiality and reveals who else might have access to customers' data, customers were rightfully concerned about their data being collected without knowing who else might view their home layout (Williamson & Prybutok, 2024).

Privacy concerns relate to informed consent regarding who collects personal data and for what purposes. Such information about data collection should be transparently disclosed to consumers, and devices should not collect any personal data without user's permission before operation (Azam et al., 2023; Grabler & Koeszegi, 2024).

Security concerns question how such data is transmitted and whether encryption methods and other cyber-security systems are in place to protect consumer data during transmissions and while stored on servers (Szynkiewicz et al., 2023; Ulsmåg et al., 2024; Wang et al., 2024). Similar issues have also been raised about other consumer robots other than vacuuming robots.

6.5 COMMON ATTACK VECTORS IN ROBOTICS

Robot systems can face a wide range of security threats due to the complexity of their hardware and software components. Attack vectors are vulnerabilities that can be exploited by attackers to gain access to a system (Bhardwaj et al., 2022). These attacks can take many forms, such as exploiting software vulnerabilities or intercepting communication signals (Botta et al., 2023; Yu et al., 2023). Understanding the different attack vectors that robotic systems can face is important to build secure systems that are resilient to attack (Shaik et al., 2025).

One attack vector is sensor spoofing, where false sensor data is fed into the system. Robots rely on sensors to perceive their environment and make decisions. If the sensor data is compromised, the robot can be tricked into behaving incorrectly (Shaik et al., 2025; Szynkiewicz et al., 2023). For instance, a robot might be made to see obstacles that are not there or miss obstacles that are.

Another attack vector is communication interception. Robotic systems often communicate with each other or with external devices wirelessly (Wu et al., 2023). If these communications are not encrypted or authenticated properly, attackers can intercept and read or modify the data being transmitted (Kosugi et al., 2023). This can lead to a breach of confidentiality or integrity depending on the data being intercepted.

System hijacking is when an attacker gains control of the robot's systems. This can happen if there are vulnerabilities in the software that controls the robot or if authentication mechanisms are weak (Zhang et al., 2022). Once an attacker has gained control of the robot, they can make it behave in any way they want.

Data leakage occurs when sensitive information is accessed by unauthorised parties. This can happen if user data is not properly protected, such as being stored in an insecure location or transmitted without encryption (Wang et al., 2024). Data leaks can lead to privacy breaches and identity theft.

These are just some examples of attack vectors that robotic systems can be vulnerable to. There are many other ways in which attackers can try to compromise robotic systems. It is important to consider all potential attack vectors when designing secure systems and implement safeguards where possible (Shaik et al., 2025; Zhang et al., 2022).

6.6 SECURE COMMUNICATION AND ENCRYPTION

Secure communications refer to methods by which two or more parties talk without risk of interception or interference from outsiders (Bhardwaj et al., 2025). Robot sensors, processors, and controllers often need to communicate over networks that can be vulnerable to attack or misuse. Encrypting messages that these components transmit to each other can help protect against these weaknesses (Nandi et al., 2023). Encryption and secure communications allow robots to protect information relating to operation and functionality.

Encryption scrambles messages to make them unreadable to anyone who does not know how to interpret the message (Tidrea et al., 2023). Encryption can protect against eavesdroppers from interpreting messages sent between components such as sensors and controllers. Symmetric encryption allows the sender and recipient of a message to use a shared key to both encode and decode information (Kovalov, 2025). A symmetric encryption algorithm can be used in tandem with cryptographic keys to keep communications secure.

Secure communication refers to the standards used to govern communications between two parties (Tidrea et al., 2023). When data is communicated securely, in addition to being encrypted, it can be guaranteed that the information has not been altered in transit. Authentication protocols can be used to verify that a communicating party is who they say they are (Zhang et al., 2022). Information can then only be accessed by verified users.

Robotics applications can utilise encryption by taking advantage of libraries that allow users to easily encrypt information. The Python programming language has a library called cryptography that allows users to easily encrypt their data (Kovalov, 2025). The Fernet Encryption allows users to both encrypt their data and authenticate their messages.

Listing 6.1: Basic encryption using Fernet in Python

```
from cryptography.fernet import Fernet

# Generate a key
key = Fernet.generate_key()
cipher = Fernet(key)

# Encrypt data
message = b"Robot sensor data"
encrypted = cipher.encrypt(message)

# Decrypt data
decrypted = cipher.decrypt(encrypted)
```

This example shows how you can encrypt and decrypt data securely inside your robot. You can use this technique to secure your communications and minimise data leakage or corruption.

Robotic systems often need to securely communicate with other devices. Encryption and authentication allow you to safely transmit information across networks while preserving privacy and integrity of your system.

6.7 PRIVACY AND COMPLIANCE IN ROBOTICS

Privacy should be considered in robotics because robots process personal data (Oruma et al., 2023). Robots collect information about users and their environment when deployed at home or workspaces and public areas, which can include habits, routines, and surrounding landscape (Callander et al., 2024).

Privacy regulations such as GDPR outline how personal data should be handled when operating in the EU (Hoxhaj et al., 2023). Transparency, collecting the least amount of data possible, and consent are very important when building systems that process personal data under GDPR. Privacy considerations in robotics draw heavily from these areas (Azam et al., 2023; Oruma et al., 2023). Robots can collect personal data continuously which users may not be aware of.

Privacy by design allows only the necessary data required for operation to be collected (Hoxhaj et al., 2023). If a robot vacuum is being used, it needs to map out its surroundings but does not need to collect information about users personally (Azam et al., 2023; Oruma et al., 2023; Ulsmåg et al., 2024).

The system should be transparent to users what data is being collected, how it is used, and who has access to it (Azam et al., 2023; Callander et al., 2024). Users should give consent to their data being collected. If not stated clearly, users will not trust the robotic system (Song et al., 2023).

Access control and methods to keep data protected from unwanted sources should be implemented. Examples of this include encryption and authentication. Privacy in robotics is important to build ethical, secure systems that users can feel comfortable with (Song et al., 2023).

6.8 PRACTICAL INSIGHT: THREAT MODELLING A HOME ROBOT

Threat modelling with robotics can be demonstrated with a robot vacuum cleaner roaming around the living room floor, creating a map of it so that it cleans the floor more effectively. It shows how threat modelling applies to both robots physically interacting with the world and processing data about it (Grabler & Koeszegi, 2024).

The first step in threat modelling is defining our assets, what do we want to protect? The assets in our robotic vacuum could include the mapping of the room, user data stored on the vacuum, commands that control the vacuum, and data sent over Wi-Fi or Bluetooth. Some specific assets that are important in robotics are the sensors and sensor data since it allows the robot to interact with the world around it (Wang et al., 2024).

Identifying potential threat actors is our next step. This includes anyone who may try to attack your system. These attackers can be anyone from someone trying to gain unauthorised access to your robot and take control of it, someone trying to intercept data being sent to/from the robot, or someone who wants to send unauthorised sensor inputs to your robot. This last example could include spoofing sensor inputs to trick your robot or intercepting and reading messages sent between your robot and your cell phone (Szynkiewicz et al., 2023).

The next step is to identify vulnerabilities within your system. This can include using weak authentication mechanisms, sending data over networks without encryption, or not validating sensor data received. Vulnerabilities allow an attacker to gain access to your assets and can let them steal sensitive data or take control of your robot (Wang et al., 2024; Yankson et al., 2023).

Identify impacts are the results if a threat were to take advantage of vulnerabilities. If someone were to gain access to the mapping data, they would be able to learn about private information about what your home looks like. If someone gained control of your robot, they could have it perform any action while it's cleaning, which could be dangerous (Grabler & Koeszegi, 2024).

By threat modelling, you can identify weaknesses before they're exploited and add the proper protections to keep robots and us safe.

6.9 CHALLENGES IN SECURING ROBOTIC SYSTEMS

Ensuring security of robotic systems can be difficult for a variety of reasons. Robot systems involve complex integration of software, hardware, and communication protocols, which can create security vulnerabilities if not properly designed and implemented (Bhardwaj et al., 2025; Haskard & Herath, 2025). Additionally, robots often operate in real-time environments and interact with the physical world, making them susceptible to unique security threats (Shaik et al., 2025).

Some challenges in securing robotic systems include system complexity. Robots can consist of multiple interconnected components such as sensors, controllers, and communication networks, which can create multiple points of potential vulnerabilities. It can be difficult to ensure security across the entire system, and all components may not be designed with consistent security in mind (Nandi et al., 2023). Real-time constraints. Many security techniques like encryption and authentication introduce additional processing overhead, which can impact system performance (Nandi et al., 2023). Robots often operate in real-time or near real-time environments where delays can cause issues.

Hardware limitations: Robots are often built on resource-constrained platforms with limited processing power and memory. It can be difficult to implement robust security measures without hindering system performance (Cagua et al., 2024).

Operating environments: Robots are often deployed in open and uncontrolled environments, making them vulnerable to a wide range of attacks (Shaik et al., 2025). Outside interference, physical tampering, and environmental factors can pose security risks.

Evolving threats: As technology continues to advance, new security threats and vulnerabilities are constantly emerging (Saeed et al., 2023). Systems must be designed to adapt and respond to evolving threats over time through updates and patches (Holdbrook et al., 2024).

Ensuring security often comes with trade-offs with other design considerations. It is important that security mechanisms do not overly degrade system performance or functionality (Bhardwaj et al., 2025). Security should be designed to fit the constraints and requirements of the system both in terms of technology and real-world environment (Haskard & Herath, 2025).

6.10 ETHICAL AND REAL-WORLD CONSIDERATIONS

Ethical and social issues: Robots interact with the physical world differently from software programs, raising ethical and social issues that must be considered (Oruma et al., 2023). Security flaws may result in data theft and loss of privacy. However, they can cause injury or damage to people and property. Developers must consider ethical concerns when creating systems that can cause harm if not managed correctly.

Responsibility: It may not be clear who is responsible if something goes wrong with a robot, such as leaking sensitive data or operating unsafely. Is it the programmer's responsibility? The manufacturer? The service provider? Concerns about responsibility must be considered when developing a security strategy (Zhang & Zhang, 2023).

Trust: Users must have faith that robots will function without risking their information. Security issues could lessen customer loyalty and prevent people from using robotic services. Developers can build trust in their products by being open about how they collect, store, and protect client data (Oruma et al., 2023; Yang et al., 2023).

Privacy and surveillance: Robots often include sensors and cameras that let them interact with their environment and collect personal data. If safeguards fail, hackers can abuse robotic systems to spy on people. Robotic networks should collect as little private data as possible and adhere to privacy rules (Oruma et al., 2023).

Security versus functionality: Every new functionality adds new risks to robotic systems. Developers must weigh the benefits of additional capabilities against safety concerns (Shaik et al., 2025).

6.11 CONCLUSION

This chapter discussed the significance of security in robotic systems, exploring how systems can safeguard data and behaviour. Key security concepts, such as the CIA triad, threat modelling, common attack vectors, and secure communication, were introduced. Real-world implications were highlighted with the case study of the Xiaomi robot vacuum incident, emphasising the consequences of inadequate privacy measures. Practical methods like encryption and threat modelling demonstrate how security considerations can be integrated into robotic systems. The challenges of system complexity, real-time constraints, evolving threats, and ethical considerations were presented. It is important to note that security is an ongoing process that needs to be regularly monitored and updated. Ethical considerations stress the responsibility of roboticists to design systems that prioritise accountability, transparency, and user trust.

6.12 DISCUSSION QUESTIONS

1. Why should cybersecurity be implemented differently in robotic systems versus traditional software systems?
2. How would you apply the CIA triad to creating secure robotic systems?

3. What threats are posed by sensor spoofing and system hijacking?
4. Why was Xiaomi's robot vacuum incident a concern for privacy/data protection?
5. Who should be responsible when robotic systems leak private information?

6.13 EXERCISES

Exercise 1: Threat Modelling Activity

Draw out your own threat model for a home robot. List at least three assets, two threats, two vulnerabilities, and impact.

Exercise 2: Identifying Attack Vectors

Pick any simple robotic system (delivery robot, smart assistant robot, etc.). Identify potential attack vectors and how to mitigate them.

Exercise 3: Secure Communication Task

Discuss how you would encrypt information passed between components in robotic systems. Provide an example with a brief description based on the Python code we used in this chapter.

Exercise 4: Reflection

If robots are collecting our data, who is responsible for protecting our privacy? How do we gain trust in these devices?

REFERENCES

Amro, A., Gkioulos, V., & Katsikas, S. (2022). Assessing cyber risk in cyber-physical systems using the ATT&CK framework. *ACM Transactions on Privacy and Security*, *25*(4), 1–38.

Azam, N., Michala, L., Ansari, S., & Truong, N. (2023). Data privacy threat modelling for autonomous systems: A survey from the GDPR's perspective. *IEEE Transactions on Big Data*, *9*(2), 244–261.

Bhardwaj, A., Alshehri, M., Kaushik, K., Alyamani, H. J., & Kumar, M. (2022). Secure framework against cyber-attacks on cyber-physical robotic systems. *Journal of Electronic Imaging*, *31*(2), 023001.

Bhardwaj, A., Bharany, S., Rehman, A., Tejani, G. G., & Hussen, S. (2025). Securing cyber-physical robotic systems for enhanced data security and real-time threat mitigation. *EURASIP Journal on Information Security*, *2025*(5), 1–26.

Botta, A., Rotbei, S., Zinno, S., & Ventre, G. (2023). Cyber security of robots: A comprehensive survey. *Intelligent Systems with Applications*, *18*, 200220.

Cagua, G., Gauthier-Umaña, V., & Lozano-Garzon, C. (2024). Implementation and performance of lightweight authentication encryption ASCON on IoT devices. *IEEE Access*, *12*, 82134–82147.

Callander, N., Ramírez-Duque, A. A., & Foster, M. E. (2024). Navigating the human-robot interaction landscape: Practical guidelines for privacy-conscious social robots. In Companion of the 2024 ACM/IEEE International Conference on Human-Robot Interaction (pp. 102–106).

Grabler, R., & Koeszegi, S. T. (2024). Privacy beyond data: Assessment and mitigation of privacy risks in robotic technology for elderly care. *ACM Transactions on Human-Robot Interaction, 13*(3), 1–35.

Haskard, A., & Herath, D. (2025). Secure robotics: Navigating challenges at the nexus of safety, trust, and cybersecurity in cyber-physical systems. *ACM Computing Surveys, 58*(5), 118.

Holdbrook, R., Odeyomi, O., Yi, S., & Roy, K. (2024). Network-based intrusion detection for industrial and robotics systems: A comprehensive survey. *Electronics, 13*(22), 4501.

Hoxhaj, O., Halilaj, B., & Harizi, A. (2023). Ethical implications and human rights violations in the age of artificial intelligence. *Balkan Social Science Review, 21*, 125–152.

Jeffrey, N., Tan, Q., & Villar, J. (2023). A review of anomaly detection strategies to detect threats to cyber-physical systems. *Electronics, 12*(15), 3217.

Jiang, Y., Wang, W., Ding, J., Lu, X., & Jing, Y. (2024). Leveraging digital twin technology for enhanced cybersecurity in cyber-physical production systems. *Future Internet, 16*(4), 125.

Kosugi, A., Teranishi, K., & Kogiso, K. (2023). Experimental validation of the attack-detection capability of encrypted control systems using man-in-the-middle attacks. *IEEE Access, 11*, 118650–118663.

Kovalov, P. (2025). Securing data exchange channels between Python applications using cryptographic libraries. *Terra Security, 11*(2), 45–60.

Meland, P. H., Nesheim, D. A., Bernsmed, K., & Sindre, G. (2022). Assessing cyber threats for storyless systems. *Journal of Information Security and Applications, 64*, 103058.

Nandi, G. S., Pereira, D., Proença, J., Tovar, E., Rodriguez, A., & Garrido, P. (2023). Secure integration of extremely resource-constrained nodes on distributed ROS2 applications. *Open Research Europe, 3*, 102.

Oruma, S. O., Ayele, Y., Sechi, F., & Rødsethol, H. (2023). Security aspects of social robots in public spaces: A systematic mapping study. *Sensors, 23*(18), 7719.

Phillips, S. C., Taylor, S., Boniface, M., Modafferi, S., & Surridge, M. (2022). Automated knowledge-based cybersecurity risk assessment of cyber-physical systems. *IEEE Access, 10*, 62341–62364.

Saeed, S., Altamimi, S. A., Alkayyal, N. A., Alshehri, E., & Alabbad, D. A. (2023). Digital transformation and cybersecurity challenges for businesses resilience: Issues and recommendations. *Sensors, 23*(15), 6641.

Shaik, A. K., Mohammadi, A., & Malik, H. A. M. (2025). A systematic review of sensor vulnerabilities and cyber-physical threats in industrial robotic systems. *IET Cyber-Physical Systems: Theory & Applications, 10*(1), 45–72.

Sheikh, Z. A., & Singh, Y. (2022). A hybrid threat assessment model for security of cyber physical systems. In 2022 Seventh International Conference on Parallel, Distributed and Grid Computing (PDGC) (pp. 303–308). IEEE.

Song, B., Xu, H., Hu, W., Li, Y., & Guo, Y. (2023). How to calculate privacy: Privacy concerns and service robots' use intention in hospitality. *Current Issues in Tourism, 27*(18), 3050–3072.

Song, X., Liang, X., & Zhou, H. (2024). Semantic mapping techniques for indoor mobile robots: Review and prospect. *Measurement and Control, 57*(9–10), 1–18.

Szynkiewicz, W., Niewiadomska-Szynkiewicz, E., & Lis, K. (2023). Deep learning of sensor data in cybersecurity of robotic systems: Overview and case study results. *Electronics, 12*(19), 4146.

Tanimu, J., & Abada, W. (2024). Addressing cybersecurity challenges in robotics: A comprehensive overview. *Cybersecurity and Applications, 4*(2), 100118.

Tidrea, A., Korodi, A., & Silea, I. (2023). Elliptic curve cryptography considerations for securing automation and SCADA systems. *Sensors, 23*(5), 2581.

Ulsmåg, B., Lin, J.-C., & Lee, M.-C. (2024). Investigating the privacy risk of using robot vacuum cleaners in smart environments. *Sensors, 24*(15), 5123.

Wang, T., Chen, K., Zheng, Z., Guo, J., Zhao, X., & Zhang, S. (2024). PrivShieldROS: An extended robot operating system integrating Ethereum and interplanetary file system for enhanced sensor data privacy. *Sensors, 24*(9), 2928.

Williamson, S. M., & Prybutok, V. R. (2024). Balancing privacy and progress: A review of privacy challenges, systemic oversight, and patient perceptions in AI-driven healthcare. *Applied Sciences, 14*(2), 705.

Wu, C., Yao, W., Luo, W., Pan, W., Sun, G., Xie, H., & Wu, L. (2023). A secure robot learning framework for cyber-attack scheduling and countermeasure. *IEEE Transactions on Robotics, 39*(5), 3590–3608.

Xing, W., & Shen, J. (2024). Security control of cyber–physical systems under cyber-attacks: A survey. *Sensors, 24*(12), 3785.

Yang, J., Chen, Y.-L., Por, L. Y., & Ku, C. S. (2023). A systematic literature review of information security in chatbots. *Applied Sciences, 13*(10), 6054.

Yankson, B., Loucks, T., Sampson, A., & Lojano, C. (2023). Robots' security assessment and analysis using open-source tools. In Proceedings of the 18th International Conference on Cyber Warfare and Security (pp. 485–494).

Yu, Z., Gao, H., Cong, X., Wu, N., & Song, H. (2023). A survey on cyber–physical systems security. *IEEE Internet of Things Journal, 10*(24), 21557–21579.

Zhang, D., Shi, P., Agarwal, R. K., & Kovács, L. (2024). Reference tracking MPC for cyber-physical systems under denial-of-service attacks: An omnidirectional robot application. *IEEE Systems Journal, 18*(2), 2441–2452.

Zhang, J., & Zhang, Z.-M. (2023). Ethics and governance of trustworthy medical artificial intelligence. *BMC Medical Informatics and Decision Making, 23*, 6.

Zhang, S., Li, W., Li, X., & Liu, B. (2022). AuthROS: Secure data sharing among robot operating systems based on Ethereum. In 2022 IEEE 22nd International Conference on Software Quality, Reliability and Security (QRS) (pp. 668–677).

Part III

Building and Controlling Robots

7 How Robots Are Built
Form, Function, and Failure

7.1 INTRODUCTION

Mechanical design dictates how a robot moves, how stable it is, how much strength it has, and how it interacts with the world around it (Gao et al., 2023). Software may tell a robot how to behave, but its mechanics determine what it can physically do (Feller, 2023). Robots are engineered to operate in real-world settings; thus, they must have mechanisms that allow them to execute actions (Fan et al., 2024). Mechanics play a major role in robotics because it defines the robot's abilities.

Robots are considered cyber-physical systems because their software, electronics, and mechanics all come together to create a functioning whole (Zahid et al., 2024). Mechanical design provides the robot with a skeleton that holds everything together and gives it the shape and form it needs to operate. Software can only tell a robot what to do, but none of its directives can be achieved if its structure cannot support it, limiting its range of performance or even breaking it outright (Aygül et al., 2025).

Mechanical design consists of choosing which mechanical components will be used in a robot, such as what joints, actuators, and chassis structure will be used. It also arranges these elements so the robot can perform its task effectively and efficiently (Feller, 2023). This requires an understanding of how the loads are placed on the robot, what range of motion is needed, and what environment the robot will be expected to work in, among other things. When designing a robot to be used in rough terrain, you would not use the same mechanical design that you would use for a small indoor rover (Gao et al., 2023).

In this chapter, we will cover the basics of robotic mechanical design. We will discuss how form can affect function and how design can limit the dependability of your system (Zahid et al., 2024). By the end of this chapter, students will better understand the importance of mechanical design in robotics (Fan et al., 2024

7.2 FOUNDATIONS OF ROBOTIC MECHANICAL DESIGN

Mechanical design elements within robotics refer to the components and structures that provide movement capabilities, stability, and interaction with the physical world (Fan et al., 2024). These foundational elements include actuators, joints, and the robot's structural design, which determine how it moves and operates (Aygül et al., 2025). A comprehensive understanding of these aspects is crucial for creating functional and efficient robots (Hamrani et al., 2024).

Actuators are devices that move or control a mechanism or system (Xavier et al., 2022). They are the components that allow robots to convert energy into motion, which can include motors, hydraulic pistons, or pneumatic systems depending on

DOI: 10.1201/9781003758501-10

the type of robot and its intended use (Shao et al., 2022). Actuators enable robots to walk, lift, rotate, and perform tasks requiring force or movement (Chen et al., 2024).

Joints are points of articulation between two parts that allow for controlled movement (Jiang et al., 2024). They are essential for determining the range of motion and flexibility of a robot. Joints can be rotational, allowing movement around an axis, or linear, permitting motion in a straight line. The robot's agility and precision are heavily dependent on the design and configuration of its joints.

The chassis or frame of a robot provides structural support and houses other critical components (Aygül et al., 2025). It is designed to offer stability and durability while often being optimised for weight to enhance the robot's efficiency and ease of movement.

Other robotic elements may include wheels, legs, or arms which aid in movement or interaction with the environment (Fan et al., 2024). Wheels enable robots to roll across surfaces efficiently, legs can allow for bipedal or quadrupedal movement suitable for uneven terrain, and arms are used for manipulation tasks (Jiang et al., 2024).

Robots operate using these elements as a system (Hamrani et al., 2024). Actuators power movement at the joints, while the robot's design determines how these parts can move and work together (Fan et al., 2024). Understanding these mechanical design elements provides insight into robot capabilities and limitations (Kim et al., 2024).

7.3 CASE STUDY: MINI CHEETAH BY MASSACHUSETTS INSTITUTE OF TECHNOLOGY

Mini Cheetah robot created by Massachusetts Institute of Technology researchers is "intended to be agile, robust and modular" (Katz et al., 2022). The robot chosen is four-legged and can demonstrate agility while performing multiple tasks (Katz et al., 2022). This example of the quadruped robot will be used throughout this chapter to highlight how design can impact performance as well as efficiency and failure. Mini Cheetah was selected primarily due to its modularity factor (Katz et al., 2022). The design of this robot allows for its legs to be separate assembled units, which makes repair and modifications easy and quick (Bledt, 2023). Each leg acts as an individual component that can be replaced in case of damage or failure.

The actuators and joints used in this robot were designed to allow for fluent movement and responsive controls (Katz et al., 2022). Due to this precise design, the robot can run, jump, and stand on its legs (Katz et al., 2022). Range of motion and joint position were carefully designed to allow for a greater degree of freedom. Lightweight materials were used and 3D printed parts were used, which helped keep the robot light while still offering support and strength where needed (Katz et al., 2022). This implies that designers put focus on reducing the weight while not compromising the strength.

Mini Cheetah's design provides an example of a trade-off. By reducing the weight of the robot, its speed is increased. However, if the strength of the material is not high enough, it may result in structural failure (Dai & Kim, 2024). Increasing the joints' mobility may decrease the stability of the robot. This quadruped is an example of when these trade-offs are weighed to meet a goal. Analysing this from the Form–Function–Failure model, the form of this object allows for fast and flexible movement. The function of this robot is to showcase agility and responsiveness to controls. While this may be, the chances of failure are low due to its design (Dai & Kim, 2024). Mini Cheetah robot shows a great example of how every design choice can affect performance, efficiency, as well as failure of the mechanism (Bledt, 2023). Designers should aim to find a good compromise between modularity, material strength, and overall performance of the system.

7.4 MECHANICAL COMPONENTS IN ROBOTICS

Robotic systems consist of various mechanical elements which influence their movement capabilities, environmental interactions, and task executions (Fan et al., 2024). Elements of robotics include actuators, joints, mobility systems, end effectors, sensors, controllers, and power supply units, among others. The selection and design of these components directly affect the robot's performance and efficiency (Ding, 2024).

Actuators are components that create movement within a robotic system (Xavier et al., 2022). They can be electric motors, hydraulic pistons, or pneumatic cylinders that convert various forms of energy into mechanical motion (Jiang et al., 2024). Electric motors are widely used for their control precision, whereas hydraulic and pneumatic systems are preferred for high-force applications (Chen et al., 2024).

Joints are mechanical connections that link two or more parts of the robot, defining their relative motion. They can be rotational or linear, offering different degrees of freedom and movement ranges (Jamil et al., 2022). Precision-engineered joints are crucial for ensuring smooth and accurate robot articulations.

Mobility elements such as wheels, legs, or tracks enable robots to move within their operating environment (Bruzzone et al., 2022). Wheeled robots excel in smooth terrains, while legged robots can handle complex, uneven surfaces. Tracks offer superior grip and are often used in rough terrain conditions (Bruzzone et al., 2022). The type of mobility system chosen is typically dependent on the robot's application and operating terrain.

Robotic arms and manipulators are designed for specific interaction tasks with objects, such as lifting, positioning, or assembly operations. These systems often incorporate several joints to provide a wide range of motion and dexterity.

The chassis or frame provides structural support for all robotic components, housing the mechanical and electronic parts securely. It should be designed to be as lightweight as possible while still maintaining structural integrity under operational loads.

Robotic systems work cohesively with these elements functioning together. The design and integration of the robot's components, such as actuators, joints, and mobility solutions, determine its functionality and range of capabilities (Fan et al., 2024). Efficient and effective robotics systems are created by understanding the role of each component and optimising them for the robot's specific application (Ding, 2024).

7.5 MATERIALS AND FABRICATION

Materials and fabrication methods determine how well a robot will perform, how long it will last, and how much it will cost to build (Bisoi et al., 2023). Mechanical components must be able to withstand forces applied during normal operation without failing or losing efficiency (Al-Sabur et al., 2025). Different materials and fabrication methods have different strengths and limitations (Sarker et al., 2024). As such, they must be chosen carefully when designing robots (Simões, 2024).

Metal, plastic, and composite materials are commonly used in robotics. Metals like aluminium and steel are strong and durable. This makes them well suited for structural components and those that must support heavier loads (Raja et al., 2024). However, metal components can add significant weight to a system. This can decrease overall efficiency and the system's mobility (Kantaros et al., 2025). Plastics are lightweight and inexpensive but may not be as strong as metal. As such, they are often not recommended for use in high-stress components. Composite materials may be used to balance strength and weight. However, they can be expensive and difficult to fabricate (Al-Sabur et al., 2025; Simões, 2024).

Additive manufacturing, more commonly known as 3D printing, has recently become a widely used fabrication method in robotics (Kantaros et al., 2025). Many robotic components can be 3D printed in a matter of hours (Bisoi et al., 2023). 3D printing can be used to easily create components customised to a specific design. It is often used in rapid prototyping to create components that can be tested before sending the final design to be manufactured traditionally. Many research machines are built using 3D printed parts.

Conversely, traditional fabrication methods such as machining and casting are typically used when producing components that require high precision or durability. Traditionally fabricated parts are usually more expensive than 3D printed parts. However, they may be required for large-scale production or situations that require a high degree of reliability (Raja et al., 2024).

Different fabrication methods can limit or allow what materials should be used in a design. A lightweight material may allow for a more mobile system but may require a fabrication method capable of producing strong parts (Kantaros et al., 2025). A durable material may be more expensive or require more complex fabrication methods (Simões, 2024).

Material choice and fabrication methods can be balanced to design an optimal system. By understanding the strengths and weaknesses of each, one can design a system that performs well and lasts a long time while not spending too much money (Bisoi et al., 2023). Robots must exist in the real world. As such, it is important to ensure that the chosen materials and fabrication methods will allow the robot to operate as intended and survive in its environment (Sarker et al., 2024).

7.6 DESIGN TRADE-OFFS IN ROBOTICS

Trade-offs in system design occur when trying to improve one factor causes another factor to become worse in robotic systems (Sathuluri et al., 2023). Some common trade-offs include balancing performance vs cost, weight vs strength, cost vs performance, flexibility vs stability, and durability vs efficiency (Abbas et al., 2024). Understanding these trade-offs is crucial for creating an efficient and practical robotic system (Sarker et al., 2024).

Weight vs. strength is a critical trade-off in robotic systems. Lightweight robots tend to have better efficiency, speed, and consume less energy, which is especially important for mobile robots. However, making a robot lighter typically means using less material, which can decrease the robot's strength and make it more prone to breaking or malfunctioning. Designers must find the right balance by choosing materials and structural designs that provide enough strength without adding unnecessary weight to the system (Silva et al., 2024).

Cost vs performance is another common consideration in robotics. Using high-quality or advanced components can significantly improve a robot's performance. Examples of these might include using a high-quality actuator or using composite materials to build the robot. However, these components can be expensive, increasing the overall cost of the robotic system (Abbas et al., 2024). On the other hand, opting for cheaper components can reduce performance but keep the system affordable.

Flexibility vs stability plays a huge role in robotics. Increasing flexibility allows the robot to manoeuvre through different environments and perform more varied tasks. Robotic arms with more joints have more degrees of freedom and can thus move in more ways. However, too much flexibility can make the robot less stable and harder to control (Al-Shuka et al., 2024). Designers must ensure that the robot will be able to perform its tasks without toppling over or being stuck in positions that render it ineffective.

Durability vs efficiency is another set of trade-offs to consider. Designing a robot to be durable will allow it to operate for extended periods without maintenance. These robots are typically built to withstand tough environments and can take a beating. Building for durability often comes at the cost of making the robot heavier or slower. Robots that are built for efficiency typically aren't as durable as their robotic counterparts.

No robot design is one-size-fits-all, every design decision made comes with trade-offs (Sathuluri et al., 2023). When designing a robotic system, it is important to consider what application it will be used for and what factors are the most important. By understanding and carefully considering these trade-offs, engineers can create well-rounded systems that perform adequately under the desired criteria (Abbas et al., 2024).

7.7 PRACTICAL INSIGHT: DESIGNING A SIMPLE ROBOT

Building a simple robot can be an ideal application to highlight mechanical design principles in an ordered manner. Mechanical design attempts to assign geometry to requirements and define assembly to reach needed behaviour (Portillo-Vélez et al., 2022). Even simple mobile robots need structure, means of locomotion, and stability prerequisites (Sucuoğlu, 2025).

Mechanical design begins by establishing goals for the robot. For instance, desired movement can range from smooth navigation indoors to off-road stability that can tackle more rugged terrain (Sebastian et al., 2025). Defining the task allows mechanical design to focus on solutions tailored for that specific purpose.

Mechanical components are then sized and assigned to support design intentions (Portillo-Vélez et al., 2022). This typically includes a chassis or frame to support the system, actuators that create movement, and wheels or tracks to enable mobility. Components can then be arrayed to improve function, such as having two wheels on each side create more stability than an offset configuration might.

Designing for weight and materials can be a great way to address efficiency and stability (Sucuoğlu, 2025). Using lightweight materials can reduce power requirements but may not support the weight of the overall system (Alshihabi et al., 2024). Weight should be distributed evenly to prevent issues with tipping or jerky movements (Sebastian et al., 2025).

Ideally, building a CAD-based sketch or mock-up can help visualise component placement (Garrosa et al., 2025). Prototyping provides a base for identifying issues or weaknesses within the mechanical design before building (Portillo-Vélez et al., 2022). Iterative development is a large part of design, ensuring each prototype more closely matches the job it needs to do (Sucuoğlu, 2025).

It should also be considered how and where the design may fail. Joints can be weak or forced to support more weight than intended (Alshihabi et al., 2024). Predicting these failures can help improve the design to account for these scenarios.

Mechanical design was shown to be applied to robotics with this simple example. Goals can be translated into physical components that, when stacked together in an effective manner, create a functional system for real-world application (Garrosa et al., 2025).

7.8 FAILURE MODES IN ROBOTIC SYSTEMS

Failures due to mechanical issues are always accounted for when designing robots. Every physical weakness has the potential to evolve into a catastrophic system failure (Andruszko & Derlukiewicz, 2024). Failure modes are essentially different types of failures that can occur while operating equipment (Tiwari et al., 2025). Listed below are several common failure modes and ways to prevent them.

Repeated mechanical stress will cause eventual failure of the material. Whether it be breaking or fracturing of that material, continuous movement will degrade its strength. Components that move repeatedly or support heavy loads are prone to failure from stress and fatigue (Andruszko & Derlukiewicz, 2024). Joints and structural frames are some of the most common places to see this.

Failure can occur in any type of joint used in a robotic system. Joints will naturally wear down or move out of place after repeated use. If there is too much friction in the joint or not enough support, the robot's range of motion can be hindered. Preventing uneven force distribution can improve the lifespan of the joints and movement of your robot.

Robot chassis and support frames can fail under mechanical stress. This usually occurs when the structure cannot withstand force from regular use (Alshihabi et al., 2024). Flaws in the design or material can lead to catastrophic structural failure of your robot (Andruszko & Derlukiewicz, 2024).

Components can become loose or misplaced if they do not line up correctly. Movable components that are misaligned can put extra stress on your robot. This can cause issues in other components due to uneven force (Galan-Uribe et al., 2023). Misalignment is usually caused by imprecise construction.

Design flaws can also cause failure in your robot. This can come from miscalculating your load capacity, movement, or workspace (Tiwari et al., 2025). The tiniest errors can spiral out of control and cause failure to your system.

Knowing common failure modes helps us prevent them from happening. By reinforcing stress points, using proper materials, and checking alignments, you can reduce chances of mechanical failure (Andruszko & Derlukiewicz, 2024). Learn more about how failure modes affect your robot design.

7.9 CHALLENGES IN MECHANICAL DESIGN

Mechanical design for real-world applications comes with additional considerations that are often overlooked for abstract problems or simulation scenarios. Engineers need to account for real-world forces, material properties, robustness, safety, and edge cases from interactions with the world or user (Bliah et al., 2025).

One example is uncertainty in the real world. Robots must deal with changing terrain, weather conditions, payloads, etc., so designs must be stable and capable of some guaranteed level of performance despite uncertain conditions (Buchner et al., 2024). Designs often need to balance multiple constraints between performance, cost, and other limits while trying to maximise some criteria such as speed, accuracy, strength, throughput, battery life, etc. (Lee et al., 2024).

Limited manufacturing processes can prevent implementation of designs that would otherwise work from a theoretical standpoint. Materials may have specific limitations, and certain features can be costly or impossible to fabricate with desired capabilities (Silva et al., 2024).

Mechanical design needs to consider interactions with other electronics, wiring, or software components. Moving parts have to avoid interference from sensors or circuitry and must be designed around how the robot is controlled (Lee et al., 2024).

Weight and size limits also need to be considered as inappropriate weight distribution can cause a robot to tip over or not fit in a desired space (Buchner et al., 2024).

Building prototypes and testing design flaws that were not revealed in theoretical design are time-consuming (Silva et al., 2024).

7.10 ETHICAL AND REAL-WORLD CONSIDERATIONS

Mechanical design also has ethical considerations as robotic engineer because robots live in our world alongside us. When software goes wrong, it may cause inconvenience, but when mechanical aspects go wrong, it can hurt people or property. Safety is always a primary concern.

User safety must be considered during the design process. Ideally, robots should be safe during partial failure. Designing for safety includes using strong materials, limiting speed, and designing robots to fail safe. Joint or actuator limits can help minimise injuries if someone gets too close.

Robotics components should be reliable as well. Mechanical elements should last a long time, and many different things can cause them to fail. Components can fail due to improper materials, weak structures, and large design margins (Aygül et al., 2025). Thoroughly stress testing your design can help prevent unexpected failures.

Ethical engineers should also consider all possible ways for something to go wrong. Testing and allowing for uncertainties in design can prevent mishaps (Sun, 2025). Adding a safety margin can account for unknowns.

Production and material choice also have some ethical responsibilities. Manufacturing your design can come with wasted material and inefficient use of resources (Giordano et al., 2023).

In general, mechanical designers should strive to design safely, reliably, and efficiently (Giordano et al., 2023).

7.11 CONCLUSION

This chapter discussed the importance of mechanical design and how it directly affects kinematics, dynamics and robotic systems usability in real-world scenarios. Throughout this chapter, it was shown that actuators, joints, and the overall mechanical structure contribute to a system's abilities, while the choice of material and fabrication technique affect a robot's longevity, efficiency and cost. The Form–Function–Failure approach provided guided steps to understand how design attributes can affect functionality and eventual failure of a mechanical design. By using the example of the Massachusetts Institute of Technology Mini Cheetah, we were able to see how modular design, choice of material and intelligent design allow for capable and fault-tolerant systems. Mechanical design trade-offs such as weight vs strength and cost vs functionality were discussed to showcase that there are many important design features that need to be considered and balanced when engineering a product. Lastly, different failure modes and real-world design considerations were presented to allow the reader to see common weaknesses in mechanical design and learn how to iteratively improve their own designs.

7.12 DISCUSSION QUESTIONS

1. How does the mechanical design of a robot impact system performance?
2. How do form, function, and failure relate to one another?
3. Why are trade-offs necessary when designing robots?
4. How can a small mistake in mechanical design cause your entire robot to fail?
5. How do ethics play into the physical design of your robot?

7.13 EXERCISES

Exercise 1: CAD Drawing

Draw (freehand) a rudimentary mobile robot as if you were drawing it in CAD. Be sure to label things like the chassis, wheels, joints, and actuators.

Exercise 2: Comparing Materials

Fill out a table comparing metals, plastics, and 3D-printed materials. You can compare strength vs. weight, cost, ease of use, or other engineering properties relevant to robot design.

Exercise 3: Identifying Failures

Pick one part of a simple robot that could fail (like a joint or wheel axle) and describe how it could fail and how that failure could be prevented.

Exercise 4: Design Decision Essay

Describe how trade-offs had to be made when designing your robot. How did your decisions impact the overall performance of your system?

REFERENCES

Abbas, M. R., Ahsan, M., & Iqbal, J. (2024). Experimental development of lightweight manipulators with improved design cycle time that leverages off-the-shelf robotic arm components. *PLOS ONE*, *19*(7), e0306719.

Al-Sabur, R., Ameen, Y. M., Khalaf, H. I., Mishra, A., & Sharkawy, A.-N. (2025). Modeling the structural dynamics of carbon fiber composites for robotic systems under sinusoidal load. *International Journal of Robotics and Control Systems*, *5*(1), 45–65.

Alshihabi, M., Ozkahraman, M., & Kayacan, M. Y. (2024). Enhancing the reliability of a robotic arm through lightweighting and vibration control with modal analysis and topology optimization. *Mechanics Based Design of Structures and Machines*, *52*(9), 2093–2114.

Al-Shuka, H., Kaleel, A. H., & Al-Bakri, B. A. R. (2024). Hierarchical stabilization and tracking control of a flexible-joint bipedal robot based on anti-windup and adaptive approximation control. *Journal of Robotics*, *2024*, 9034218.

Andruszko, J., & Derlukiewicz, D. (2024). Enhancing demolition works safety: Integrating numerical and experimental methods for structural failure prevention in remote-controlled demolition robots. *Eksploatacja i Niezawodność – Maintenance and Reliability*, 26(4), 510–520.

Aygül, C., Güven, C., Frunzi, S. A., Katz, B. J., & Nemitz, M. P. (2025). A framework for soft mechanism driven robots. *Nature Communications*, *16*(1), 1032.

Bisoi, A., Tüfekci, M., Öztekin, V., Denimal Goy, E., & Salles, L. (2023). Experimental investigation of mechanical properties of additively manufactured fibre-reinforced composite structures for robotic applications. *Applied Composite Materials*, *30*(6), 3499–3525.

Bledt, G. (2023). Advances in modular legged robot design for dynamic locomotion. *IEEE Robotics & Automation Magazine*, *30*(1), 52–64.

Bliah, O., Hegde, C., Tan, J. M. R., & Magdassi, S. (2025). Fabrication of soft robotics by additive manufacturing: From materials to applications. *Chemical Reviews*, *125*(18), 11021–11069.

Bruzzone, L., Nodehi, S. E., & Fanghella, P. (2022). Tracked locomotion systems for ground mobile robots: A review. *Machines*, *10*(8), 667.

Buchner, T. J. K., Fukushima, T., Kazemipour, A., Gravert, S.-D., Prairie, M., Romanescu, P., Arm, P., Zhang, Y., Wang, X., Zhang, S. L., Walter, J. R., Keplinger, C., & Katzschmann, R. K. (2024). Electrohydraulic musculoskeletal robotic leg for agile, adaptive, yet energy-efficient locomotion. *Nature Communications*, *15*, 6214.

Chen, X., Ma, G., Brugnera, A., Murcia, M., Mohseni, O., Takahashi, R., Forner-Cordero, A., Seyfarth, A., Hosoda, K., & Sharbafi, M. A. (2024). Design of low-cost modular bio-inspired electric–pneumatic actuator-driven legged robots. *Biomimetics*, 9(2), 123.

Dai, H., & Kim, S. (2024). Structural optimization and material selection for agile legged robots. *International Journal of Robotics Research*, *43*(2–3), 210–229.

Ding, F. (2024). Finite element analysis and optimal design of robotic arm. *Applied and Computational Engineering*, *4*(1), 15–28.

Fan, Y., Pei, Z., Wang, C., Li, M., Tang, Z., & Liu, Q. (2024). A review of quadruped robots: Structure, control, and autonomous motion. *Advanced Intelligent Systems*, *6*(8), 2300456.

Feller, D. (2023). Dexterity, workspace and performance analysis of the conceptual design of a novel three-legged, redundant, lightweight, compliant, serial-parallel robot. *Journal of Intelligent & Robotic Systems*, *108*, 35.

Galan-Uribe, E., Morales-Velazquez, L., & Osornio-Ríos, R. (2023). FPGA-based methodology for detecting positional accuracy degradation in industrial robots. *Applied Sciences*, *13*(15), 8874.

Gao, X., Guan, X., Yang, Y., & Zhang, J. (2023). Design and ground performance evaluation of a multi-joint wheel-track composite mobile robot for enhanced terrain adaptability. *Applied Sciences*, *13*(12), 7045.

Garrosa, M., Ceccarelli, M., Russo, M., & Yang, B. (2025). Design of a mobile assisting robot for blind and elderly people. *Applied Sciences*, *15*(10), 5120.

Giordano, G., Babu, S. P. M., & Mazzolai, B. (2023). Soft robotics towards sustainable development goals and climate actions. *Frontiers in Robotics and AI*, *10*, 1122334.

Hamrani, A., Rayhan, M. M., Mackenson, T., McDaniel, D., & Lagos, L. (2024). Smart quadruped robotics: A systematic review of design, control, sensing and perception. *Advanced Robotics*, *38*(19), 1105–1134.

Jamil, B., Rodrigue, H., & Choi, Y. (2022). Design of a novel sensing method for a pneumatic artificial muscle actuator-driven 2-degrees of freedom parallel joint. *Soft Robotics*, *9*(3), 493–505.

Jiang, Z., Wang, Y., & Zhang, K. (2024). Development of a pneumatically actuated quadruped robot using soft–rigid hybrid rotary joints. *Robotics*, *13*(1), 9.

Kantaros, A., Drosos, C., Papoutsidakis, M., Pallis, E., & Ganetsos, T. (2025). Composite filament materials for 3D-printed drone parts: Advancements in mechanical strength, weight optimization and embedded electronics. *Materials*, *18*(11), 4012.

Katz, B., Chignoli, M., & Kim, S. (2022). Mini Cheetah: A platform for exploring dynamic quadrupedal locomotion. *Robotics and Autonomous Systems*, *154*, 104130.

Kim, T., Kaarthik, P., & Truby, R. L. (2024). A flexible, architected soft robotic actuator for motorized extensional motion. *Advanced Intelligent Systems*, *6*(15), 2300866.

Lee, J., Bjelonic, M., Reske, A., Wellhausen, L., Miki, T., & Hutter, M. (2024). Learning robust autonomous navigation and locomotion for wheeled-legged robots. *Science Robotics*, *9*(89), eadi9340.

Portillo-Vélez, R. D. J., Burgos-Castro, I. A., Vásquez-Santacruz, J., & Marín-Urías, L. et al (2022). Integrated conceptual mechatronic design of a Delta robot. *Machines*, *10*(3), 214.

Raja, S., AhmedMustafa, M., Ghadir, G. K., Al-Tmimi, H. M., Alani, Z. K., Rusho, M. A., & N, R. (2024). An analysis of polymer material selection and design optimization to improve structural integrity in 3D printed aerospace components. *Applied Chemical Engineering*, *8*(1), 55–70.

Sarker, A., Islam, T. U., & Islam, M. R. (2024). A review on recent trends of bioinspired soft robotics: Actuators, control methods, materials selection, sensors, challenges, and future prospects. *Advanced Intelligent Systems*, *6*(10), 2300321.

Sathuluri, A., Sureshbabu, A. V., Frank, J., Amm, M., & Zimmermann, M. (2023). Computational systems design of low-cost lightweight robots. *Robotics*, *12*(4), 91.

Sebastian, A., Murali, V., Pragna, R., & Aklak, A. F. (2025). Design of capsule-shaped all-terrain robot: A perpendicular track mobility solution. *IEEE Access*, *13*, 45678–45692.

Shao, J., Bian, Y., Yang, M., & Liu, G. (2022). Characteristic analysis and motion control of a novel ball double-screw hydraulic robot joint. *Engineering Applications of Computational Fluid Mechanics*, *16*(1), 1065–1085.

Silva, A., Fonseca, D., Neto, D., Babcinschi, M., & Neto, P. (2024). Integrated design and fabrication of pneumatic soft robot actuators in a single casting step. *Cyborg and Bionic Systems*, *5*, 0094.

Simões, S. (2024). High-performance advanced composites in multifunctional material design: State of the art, challenges, and future directions. *Materials*, *17*(22), 7123.

Sucuoğlu, H. S. (2025). Development of topologically optimized mobile robotic system with machine learning-based energy-efficient path planning structure. *Machines*, *13*(7), 845.

Sun, J. (2025). Human failure analysis and fault-tolerant interaction design of deep-sea exploration robots. *International Scientific Technical and Economic Research*, *5*(3), 88–104.

Tiwari, A., Kumar, S., Sharma, R., Mehdi, H., & Saroha, M. (2025). Analysing the reliability factors of a robot utilized within an FMC comprising two machines and one robot. *International Journal on Interactive Design and Manufacturing*, *19*(1), 123–138.

Xavier, M. S., Tawk, C. D., Zolfagharian, A., Pinskier, J., Howard, D., Young, T. R., Lai, J., Harrison, S. M., Yong, Y., Bodaghi, M., & Fleming, A. J. (2022). Soft pneumatic actuators: A review of design, fabrication, modeling, sensing, control and applications. *IEEE Access*, *10*, 5111–5158.

Zahid, M., Bucaioni, A., & Flammini, F. (2024). Model-based trustworthiness evaluation of autonomous cyber-physical production systems: A systematic mapping study. *ACM Computing Surveys*, *56*(2), 39.

8 Control and Feedback in Robotics

8.1 INTRODUCTION

When building robots and programs to control them, it's not enough to just understand their structure or how to write code that makes them move around. We also need ways of causing robots to behave in a way that we want them to, accurately and stably (Jiang et al., 2022). Previous chapters discussed robot structures and programming robots. In this chapter, we discuss controlling robots.

Control allows a system to change its behaviour based on its current situation (Cvitanić & Melkote, 2022). We use control to make systems behave the way we want them to. If a robot didn't use control (i.e. there was no feedback), then it would be flying blind. Robot arms don't simply move to a position and stay there. They constantly monitor and adjust their motion until the arm reaches its target location (Zakeri & Xie, 2024).

Systems requiring accurate, stable, and responsive behaviour use control. This includes industrial robots, navigation, and robot arms manipulating objects. Noise in control systems can cause large errors in these applications. Feedback allows the detection of these errors so that the system can react while the error is still present (Khan, 2024).

This chapter will cover how feedback and control work. Students will learn how information about the state of a system can be used to control it (Kondratenko et al., 2022). Topics will include different types of control systems, how sensors are used to provide feedback, and PID control using mathematical models of the system you are controlling. By the end of this chapter, students will understand how robots are able to perform tasks reliably and with high accuracy.

8.2 FOUNDATIONS OF CONTROL SYSTEMS

Control systems dictate how robots behave (Jiang et al., 2022). This means that they specify the relationship between input and output. This could include response times, overshoot, accuracy, stability, and much more (Li et al., 2024). Control systems allow us to ensure robots behave as expected, especially when interacting with external objects and environments that may be uncertain or unstable.

Control can either be open loop or closed loop. Open loop control means that a system behaves solely based on the input that is provided. For instance, if we command a robot to drive forwards for 3 seconds, we will never know if it drives far enough to reach its destination. It just drives for time specified and then stops. Open loop controllers are fast and efficient but cannot adjust for any errors.

DOI: 10.1201/9781003758501-11

However, a closed-loop control system will always check where it is. That is, with some sort of sensor, the system can know its current position. This can be compared to a desired position, often called the setpoint. The difference between where the system is and where it wants to be is known as the error. We can then use this error to adjust how the system behaves to reduce this error (Li et al., 2024).

Closed loop control is implemented in many robots because it allows the robot to know where it is always. Once again, consider the robotic arm; we can have a sensor that knows the current angle. If the robot was to encounter some sort of error that caused the arm to move, it would be corrected by the closed-loop control system (Nguyet & Ba, 2023). The ideal response from a control system is one which minimises error quickly and efficiently without causing instability or oscillations (Sun et al., 2023). Knowing these basics of control will allow us to better understand more complex control methods such as PID control later in the chapter.

8.3 MODELLING FEEDBACK IN ROBOTIC SYSTEMS

Feedback loops are commonly used when representing control systems because it allows engineers to easily visualise the flow of information throughout their system. Feedback loop models show how error signals are used to adjust outputs throughout the control system (Budin & Osman, 2022). Feedback loops are essential for closed-loop control systems (Cvitanić & Melkote, 2022). The controller defines the behaviour of the system for a particular input. It takes in the desired value for the system, which is commonly known as the setpoint. The controller also calculates error between this desired value and the current state of the system (Budin & Osman, 2022). After taking these inputs, the controller will output commands to the plant. The plant represents the robot or subsystem itself. This might be an individual motor, joint, valve, etc. Finally, whatever action is sent to the plant will produce an output such as velocity or positional data (Ramírez-Neria et al., 2023). However, for a controller to use feedback to ensure accurate control of the system, sensors will take readings of the output so it can inform how to update the error moving forward (Li & Xu, 2025). Systems can react to these errors, and because of this, they can be affected by disturbances to the system (Hanson et al., 2024).

Figure 8.1 presents this loop-feedback sequence diagrammatically. Note that the process repeats indefinitely. By constantly comparing the desired setpoint to the measured output of the system, determining the error and having the controller manipulate the system as needed, steady-state or controlled operation can be maintained.

Feedback loops stress the necessity of information passing through the control mechanism. In robot motion controllers, behaviour is continuously gauged against some standard, and necessary corrections are made to achieve or maintain the desired performance. Feedback loops also form the basis for more sophisticated control schemes, such as PID controllers, which allow further tuning of error-management behaviour.

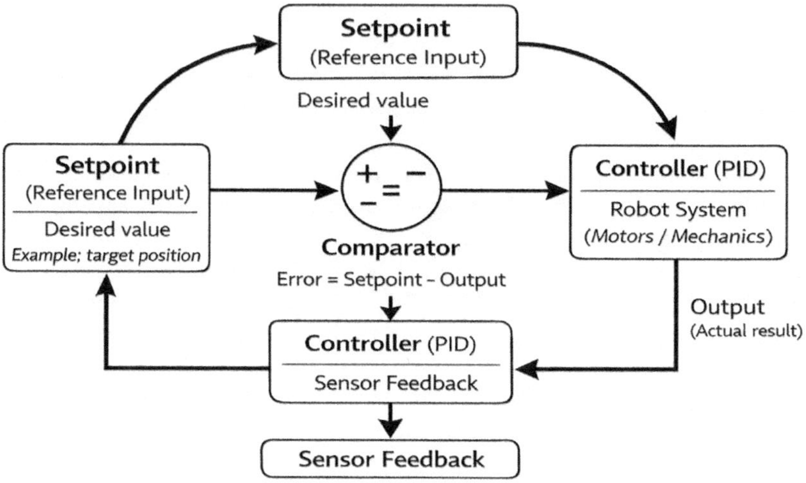

FIGURE 8.1 Closed-loop feedback system in robotics.

8.4 CASE STUDY: FEEDBACK CONTROL IN INDUSTRIAL ROBOTIC ARMS

Industrial robotic arms are mechanical devices that can be programmed to perform a variety of tasks with precision, stability, and safety (Li et al., 2023). They are commonly used in manufacturing processes such as assembly, welding, and material handling due to their repeatability and accuracy (Ārents & Greitans, 2022). The motion of a robotic arm is achieved by actuating its joints and linkages to reach a desired position or perform a task. However, various factors such as mechanical backlash, external disturbances, and payload variations can cause errors in the motion trajectory. To correct these errors and ensure precise movement, robotic arms utilise closed-loop control systems (Cvitanić & Melkote, 2022; Karlova-Sergieva et al., 2026). Closed-loop control systems use feedback sensors to measure the position, velocity, force, or torque of the robotic arm (Li et al., 2023; Lin et al., 2023). The feedback information is compared with the desired setpoint or reference value to generate an error signal. The controller then adjusts the control signals sent to the actuators based on this error signal, which corrects the arm's movement. For instance, when the arm is commanded to move to a specific position, the feedback sensors continuously monitor whether it has reached that position (Li et al., 2023). If the arm is slightly off target, the controller generates corrective actions to bring it closer to the desired position. This error correction happens in real-time, allowing the robotic arm to reach the desired position with high accuracy (Li et al., 2023).

Feedback control is also used to maintain stability and safety of the robotic arm (Cvitanić & Melkote, 2022; Karlova-Sergieva et al., 2026). For example, when lifting heavy payloads or moving at high speeds, even small errors can result in instability or collisions. Feedback control helps to limit the movement of the arm, reduce oscillations, and ensure smooth motion (Li et al., 2023). Industrial robotic arms illustrate the effectiveness of the feedback loop model from a systems perspective (Khan, 2024). Sensors provide real-time feedback to the controller, which then adjusts the control signals sent to the actuators (Kunal et al., 2025). This continuous cycle of measurement, comparison, and correction enables the robotic arm to perform tasks accurately and reliably. In conclusion, industrial robotic arms demonstrate how feedback control systems can be applied to solve real-world problems (Ārents & Greitans, 2022). Without feedback, control systems would not be able to adapt to changing environments, resulting in poor performance and higher risks of failure (Li et al., 2023).

8.5 PID CONTROL IN ROBOTICS

PID control stands for Proportional-Integral-Derivative control and is a common technique used in robotics to control feedback systems (Zhang et al., 2025). PID control allows for stable, accurate control of a system by minimising error using three separate controls (Kong et al., 2024). Each individual component affects the robot in different ways. The proportional control acts based on the current value of the error. This value is also known as the setpoint (Zhang et al., 2025). A larger proportional value will result in a more significant change as a reaction to an error. If only proportional control is used, the system will typically result in a small amount of error (Ortatepe, 2023).

Integral control considers the error over time and corrects for it (Kong et al., 2024). It will continue to push the system towards the setpoint until the error is equal to 0 (Zhang et al., 2025). Too much integral value can cause the system to become out of control. Derivative control allows the system to understand how the error is changing at a particular moment in time (Zhang et al., 2025). Based on its understanding of the system, it can account for this change and reduce overshoot (Inah et al., 2025). Derivative control allows the system to recognise the trend of the error and smooth out its response accordingly (Yapp et al., 2024).

All three parts of the PID controller work together to provide a balance of responsiveness, stability, and accuracy (Kong et al., 2024). Proportional allows the system to react, integral allows it to know how far away from the setpoint it is and derivative slows it down (Zhang et al., 2025).

The equation for a PID controller is:

$$u(t) = K_p e(t) + K_i \int e(t)dt + K_d \frac{de(t)}{dt}$$

where $u(t)$ is the control output, $e(t)$ is the error, and K_p, K_i, and K_d are the tuning parameters for each respective component.

PID control is often employed in robotic systems to control motors, track positions, and maintain balance. By tweaking the values of the proportional, integral, and derivative gains, one can adjust the system's responsiveness and stability.

PID controllers are commonly used in robotics for tasks like motor control, position control, and balancing robots. The proportional, integral, and derivative gains can be tuned to achieve the desired system response.

8.6 SYSTEM BEHAVIOUR AND ERROR CORRECTION

When evaluating system response, it's important to pay attention to how quickly and effectively it reacts to error signals over time (Joseph et al., 2022). This dynamic performance can directly impact the effectiveness of robotic systems (Liu et al., 2024). Some key aspects of system behaviour include stability, accuracy, and performance itself (Nguyet & Ba, 2023). By knowing these factors well, you'll be able to better analyse and improve your control method. Let's discuss what types of responses you should expect when using PID control. If your system overshoots its setpoint, it means that it exceeds the desired value during its response (Joseph et al., 2022). Think of a robotic arm swinging towards a target position (Ghani et al., 2023). As it reaches the setpoint, it may move slightly past it before settling. A certain amount of overshoot is normal in most systems. However, too much overshoot can cause problems. Your system may become unstable or even damage itself if it continues to exceed setpoints by large amounts (Lin et al., 2022).

Steady-state error is the amount of error that remains after a system has settled (Joseph et al., 2022). Ideally, you want this amount to be zero or as close to it as possible (Zhai & Xia, 2022). The integral term in the PID equation is what helps eliminate steady-state error (Joseph et al., 2022). Oscillation refers to any periodic movement around the setpoint (Lin et al., 2022). Suppose you set a robotic arm to a fixed position. If the parameters are not tuned correctly, it may move back and forth around the setpoint. This is because the system continuously overcorrects itself (Joseph et al., 2022). Oscillation wastes energy and can also lead to instability in some systems.

Your control system should always be stable (Vo et al., 2022). Stability refers to how your system reacts to disturbances. If a system is stable, it will return to its original state after being disturbed. An unstable system continues to oscillate or drift further away from the setpoint. When correcting errors, you never want your system to be late to the party (Lin et al., 2022). This means that your system should respond quickly to error, have little to no overshoot, eliminate steady-state error, and be stable (Joseph et al., 2022). You can achieve this by tuning your controller parameters and knowing how each affects system behaviour. Examining your system's response will allow you to create better control systems for your robots. This ensures that they'll operate smoothly no matter what the environment throws at them (Eltayeb et al., 2024; Vo et al., 2022).

8.7 SENSOR INTEGRATION IN CONTROL SYSTEMS

Sensors allow feedback control systems to track and adapt to disturbances by providing real-time measurements of the system's performance (Jain, 2025). They are required for a controller to have any idea of what the system is doing at the current time. If there are no sensors provided to give feedback on the system's output, there is no way for the controller to know if the control goal is being met and therefore no way to correct the system based on that error signal. Sensors allow a robotic system to be precise, stable, and adaptable to changing conditions (Kunal et al., 2025).

Feedback control is typically closed loop, using sensor measurements of one or more output variables of interest, such as position, velocity, temperature, or force

(Hegde et al., 2023). The sensor measures the true value of the system's output and feeds it back to the controller. The error is computed by subtracting the measurement from the setpoint. The control action is then based on this error signal (Karlova-Sergieva et al., 2026). Position sensors allow the robotic system to know where it is; without that feedback, the controller would have no reference of where the robotic arm thinks it is pointing (Zakeri & Xie, 2024).

Control system performance relies on accurate sensor feedback (Jain, 2025). The quality of measurements that the sensors receive will determine how well your system can control the process. If your sensor provides incorrect information about the process, whether too high or too low, then your controller will try to counteract that incorrect information. This will leave your system offset from its original setpoint or oscillating around that point (Ma et al., 2024).

Sensor noise is the background randomness or fuzz introduced into your measurement by the sensor. Noise will create uncertainty in the measurement data and may cause your measurement to jump around. Filtering can be used to help minimise noise (Liu et al., 2023).

Latency is the time delay between when a sensor receives a signal and when it communicates that signal to the controller. Control systems must deal with latency because they are typically real-time systems concerned with how fast the system can respond to disturbances (Jain, 2025). Having a large delay in the feedback loop will cause your controller to overshoot the setpoint because it will be reacting to old information (Kuck & Sand, 2024). Latency can be minimised with proper sensor selection and placement.

Sensor implementation involves selecting and placing the appropriate sensors and processing the sensor signals (Ma et al., 2024). Ensuring that your system has feedback allows for a control system to maintain a stable and precise system.

8.8 PRACTICAL INSIGHT: SIMULATING PID CONTROL

Simulations allow students to see how changing feedback and PID parameters affect the system behaviour (Nguyet & Ba, 2023). Systems can be modelled using software that mimics how the system will respond to certain control inputs. No physical system is necessary to make these demonstrations, which allows for great flexibility in teaching concepts (Dharmasiri & Jayananda, 2024). For this reason, simulating control systems is an effective method to demonstrate how changing PID parameters will affect system behaviour such as overshoot, stability, and steady-state error (Aner et al., 2024).

One of the simplest simulations you can perform is modelling a system that attempts to change a variable over time to a setpoint value (Dharmasiri & Jayananda, 2024). The PID will generate a control signal based on the proportional, integral, and derivative gains and the current error, cumulative error, and rate error, respectively. By changing these three gains, students can see how the system reacts to these different parameters.

If the proportional gain is too high, the system may run to the setpoint very quickly but overshoot. If the integral gain is too high, the system may start to oscillate. The derivative action will dampen this response to help reduce overshoot and allow the system to settle at the setpoint faster (Dharmasiri & Jayananda, 2024). An example using Python can be seen below.

Listing 8.1: PID control simulation in Python

```
import numpy as np
import matplotlib.pyplot as plt

# Time setup
time = np.linspace(0, 10, 100)
setpoint = 1.0

# PID parameters
Kp = 1.2
Ki = 0.5
Kd = 0.2

# Initial values
error_sum = 0
previous_error = 0
output = []

current_value = 0

for t in time:
    error = setpoint - current_value
    error_sum += error
    d_error = error - previous_error

    control = Kp*error + Ki*error_sum + Kd*d_error
    current_value += control * 0.1  # system response

    output.append(current_value)
    previous_error = error

# Plot
plt.plot(time, output, label="System Output")
plt.axhline(setpoint, linestyle='--', label="Setpoint")
plt.legend()
plt.xlabel("Time")
plt.ylabel("Value")
plt.title("PID Control Simulation")
plt.show()
```

This simulation allows students to see how the system reaches the setpoint as time progresses. Students can alter the PID constants to see different behaviours and remember the theory we talked about earlier in the chapter. The simulation allows students to gain an intuition about how controls work and are easier to apply to real-world robotics.

8.9 CHALLENGES IN CONTROL SYSTEMS

There are a few complications when choosing an adequate control design for actual robotic systems.

Control parameters will need to be tuned, and sometimes trial and error can be difficult with PID gains. Improper tuning can lead to overshoot, slow response times, and even instability (Kuck & Sands, 2024). Simulation, as well as physical experimentation, is required when tuning parameters.

Noise and inaccuracies from sensors can also lead to decreased performance. Unwanted or incorrect oscillations can occur if the feedback values being used to calculate the control signal are off. Filtered noise can improve this but can cause other issues such as decreased responsiveness (Kuck & Sands, 2024).

Latency in the feedback loop is undesirable. When commands are sent from the sensors to the actuators, there will always be some small amount of delay. The larger the delay, the more chance there is of instability due to the controller using stale values to create control signals (Lei & Song, 2024).

Systems are also often nonlinear in many aspects, which makes control design more difficult. Robots will rarely ever respond linearly to input, which limits the usage of some controllers.

Performance can be affected by external forces or disturbances. Changes in system loading, terrain, gravity, etc. can cause instability if the controller is not robust enough (Zhang et al., 2025).

Theory only goes so far in control, and there will usually be a lot of iteration when testing different controllers.

8.10 ETHICAL AND REAL-WORLD CONSIDERATIONS

Control systems directly impact the behaviour and motion of robots. This places ethical responsibility on control engineers and designers to ensure systems perform safely and as intended. Poor control can cause unsafe motion, system instability, or even damage to parts of the robot interacting with the environment or humans (Azimi & Hutchinson, 2022).

Ensuring safety should be a primary concern. An improperly tuned controller can cause large motions, oscillations, and system instability. This can damage parts, tools, or worse, cause injury to human operators. In industrial environments, this may cause parts to break or chop off fingers. Ensuring you have chosen the right type of control strategy and that the system operates stably is one way to ensure safety.

Unsafe behaviour can arise from overly aggressive controller behaviour. Trying to correct for an error too hard can cause instability. This becomes dangerous when the system is running autonomously with some input continually telling the robot to move (Azimi & Hutchinson, 2022). Ensuring proper tuning and validation can prevent this type of catastrophic divergence.

Responsibility, reliability, and robustness are also key to control systems. If your controller only works well for a select number of conditions, it will be hard to trust. Conditions such as varying payload and disturbances should be taken into consideration when validating your control system. If your system doesn't perform consistently, it will be unsafe to use in real-world applications.

Lastly, ethical responsibility and accountability fall upon engineers and designers to ensure systems perform safely. This can be done by testing, implementing safety measures, and designing for fault conditions. Control systems should always ensure safety and reliability to reduce the risk to humans in real-world applications.

8.11 CONCLUSION

This chapter discussed the use of control and feedback to allow robots to behave exactly as intended with consistency, stability, and flexibility. The concepts of open-loop versus closed-loop control were presented, demonstrating how feedback allows systems to respond to errors and maintain the desired behaviour now it matters.

Feedback loop modelling allowed for a general template to analyse information flow through a control system. An example control system using industrial robotic arms was then used to connect these concepts to physical hardware. The general topic of PID control was then discussed to control error using proportional, integral, and derivative response. These topics were combined to show how to analyse system behaviour such as overshoot, steady-state error, and stability. Sensors were discussed to allow for feedback that is both accurate and timely, along with methods to incorporate sensor readings.

Topics such as parameter tuning, noise, delays, and outside disturbances were presented as real-world complications that must be considered. Ethical concerns were also discussed to reiterate the importance of safe and reliable control. In conclusion, this chapter gave insight into how robots are able to know where they are and how they are moving to allow for exact behaviour every time.

8.12 DISCUSSION QUESTIONS

1. Why do we need feedback for precise control?
2. What is the difference between an open-loop and closed-loop control?
3. How do the PID components affect system response?
4. What creates overshoot and instability?
5. Why do you need accurate sensors for feedback control?

8.13 EXERCISES

Exercise 1: PID Tuning Simulation

Use a simulation to tune P, I, and D values. See how changing each parameter affects system response such as overshoot and stability.

Exercise 2: Open vs Closed Loop Systems

Give examples of open loop and closed loop controls used in robotics. What are some pros/cons of each?

Exercise 3: Analysing a System Response

Given a graph of a system response, identify and describe features such as rise time, overshoot, and steady state error.

Exercise 4: Reflection on the Use of Control Systems

Why are control systems necessary for safe and reliable robot operation? Provide real world examples.

REFERENCES

Aner, E. A., Awad, M. I., & Shehata, O. M. (2024). Performance evaluation of PSO-PID and PSO-FLC for continuum robot's developed modeling and control. *Scientific Reports*, *14*, 8765.

Ārents, J., & Greitans, M. (2022). Smart industrial robot control trends, challenges and opportunities within manufacturing. *Applied Sciences*, *12*(3), 1150.

Azimi, V., & Hutchinson, S. (2022). Robust adaptive control Lyapunov–barrier function for non-collocated control and safety of underactuated robotic systems. *International Journal of Robust and Nonlinear Control*, *32*(9), 5277–5305.

Budin, N. S. E., & Osman, K. (2022). Modelling and proportional-integral-derivative controller design for position analysis of a 3-degree-of-freedom robot arm. *Indonesian Journal of Electrical Engineering and Computer Science*, *27*(1), 356–364.

Cvitanić, T., & Melkote, S. N. (2022). A new method for closed-loop stability prediction in industrial robots. *Robotics and Computer-Integrated Manufacturing*, *74*, 102270.

Dharmasiri, R. A. D. D., & Jayananda, M. K. (2024). A study of two-wheel self-balancing robot dynamics for control systems learning. *Vidyodaya Journal of Science*, *29*(2), 79–96.

Eltayeb, A., Ahmed, G., Imran, I. H., Alyazidi, N. M., & Abubaker, A. (2024). Comparative analysis: Fractional PID vs. PID controllers for robotic arm using genetic algorithm optimization. *Automation*, *5*(3), 46.

Ghani, N. M. A., Othman, A., Hashim, A. A. A., & Nasir, A. N. K. (2023). Comparative analysis of PID and fuzzy logic controllers for position control in double-link robotic manipulators. *Journal of Intelligent Systems and Control*, *5*(4), 233–248.

Hanson, N., Mensah, I. A., Roberts, S. F., Healey, J., Wu, C., & Dorsey, K. L. (2024). Controlling the fold: Proprioceptive feedback in a soft origami robot. *Frontiers in Robotics and AI*, *11*, 1356572.

Hegde, C., Su, J., Tan, J. M. R., He, K., Chen, X., & Magdassi, S. (2023). Sensing in soft robotics. *ACS Nano*, *17*(15), 14463–14488.

Inah, O. M., Nazif, D. M., Umar, S., Muhammad, F., & Bal, Y. B. (2025). Performance comparison of tilt integral derivative (TID) controller and proportional integral derivative (PID) controller for parabolic dish antenna system. *Mikailalsys Journal of Advanced Engineering International*, 2(2), 45–60.

Jain, K. S. (2025). The rise of sensors for robotics in real-world applications: A technical review. *European Journal of Computer Science and Information Technology*, *13*(36), 30–41.

Jiang, J., Yang, Z., Ferreira, A., & Zhang, L. (2022). Control and autonomy of microrobots: Recent progress and perspective. *Advanced Intelligent Systems*, *4*(7), 2200030.

Joseph, S. B., Dada, E. G., Abidemi, A., Oyewola, D. O., & Khammas, B. M. (2022). Metaheuristic algorithms for PID controller parameters tuning: Review, approaches and open problems. *Heliyon*, *8*(5), e09372.

Karlova-Sergieva, V., Grasiani, B., & Nikolova, N. (2026). An integrated cyber-physical digital twin architecture with quantitative feedback theory robust control for NIS2-aligned industrial robotics. *Sensors*, *26*(2), 512.

Khan, G. D. (2024). Control of robot manipulators with uncertain closed architecture using neural networks. *Intelligent Service Robotics*, *17*(1), 55–70.

Kondratenko, Y., Atamanyuk, I., Sidenko, I., Kondratenko, G., & Sichevskyi, S. (2022). Machine learning techniques for increasing efficiency of the robot's sensor and control information processing. *Sensors*, *22*(3), 1045.

Kong, W., Zhang, H., Yang, X., Yao, Z., Wang, R., Yang, W., & Zhang, J. (2024). PID control algorithm based on multistrategy enhanced dung beetle optimizer and back propagation neural network for DC motor control. *Scientific Reports*, *14*, 12345.

Kuck, E., & Sands, T. A. (2024). Space robot sensor noise amelioration using trajectory shaping. *Sensors*, *24*(1), 123.

Kunal, K., Kathiravan, M., Madeshwaren, V., Chandrakala, T., Ganesan, V., & Gupta, S. (2025). AI-driven intelligent control strategies for industrial robotics: A reinforcement learning approach. *International Journal of Basic and Applied Sciences*, *14*(2), 101–120.

Lei, J., & Song, J.-Q. (2024). Output feedback linearization control for nanorobots with actuator and sensor delays. *Ferroelectrics*, *610*(1), 45–62.

Li, R., Zhao, Y., & Liu, H. (2023). Real-time trajectory position error compensation technology of industrial robot. *Measurement*, *205*, 112144.

Li, S., & Xu, J. (2025). Multiaxis force/torque sensor technologies: Design principles and robotic force control applications—A review. *IEEE Sensors Journal*, *25*(3), 1234–1259.

Li, T., Zhang, G., Zhang, T., & Pan, J. (2024). Adaptive neural network tracking control of robotic manipulators based on disturbance observer. *Processes*, *12*(3), 543.

Lin, C.-Y., Tran, C., Shah, S., & Ahmad, A. R. (2023). Real-time robot pose correction on curved surface employing 6-axis force/torque sensor. *IEEE Access*, *11*, 45012–45026.

Lin, J., Zheng, R., Zhang, Y., Feng, J., Li, W., & Luo, K. (2022). CFHBA-PID algorithm: Dual-loop PID balancing robot attitude control algorithm based on complementary factor and honey badger algorithm. *Sensors*, *22*(11), 4123.

Liu, C., Li, J., He, Y., Jing, A., & Li, L. (2024). Neuroadaptive-based fixed-time control for robotic manipulators with uniform prescribed performance under unknown disturbance. *International Journal of Robust and Nonlinear Control*, *34*(11), 6405–6427.

Liu, X., Clark, G., Campbell, J., Zhou, Y., & Amor, H. B. (2023). Enhancing state estimation in robots: A data-driven approach with differentiable ensemble Kalman filters. In 2023 IEEE/RSJ International Conference on Intelligent Robots and Systems (pp. 12034–12041). IEEE.

Ma, J., Li, J., Yang, Y., Hu, W., Zhang, L., & Liu, Z. (2024). Multi-sensor fusion for state estimation and control of cable-driven soft robots. *Journal of Bionic Engineering*, *21*(5), 1023–1037.

Nguyet, N. T. M., & Ba, D. X. (2023). A neural flexible PID controller for task-space control of robotic manipulators. *Frontiers in Robotics and AI*, *10*, 1122339.

Ortatepe, Z. (2023). Genetic algorithm based PID tuning software design and implementation for a DC motor control system. *Gazi University Journal of Science Part A: Engineering and Innovation*, *10*(3), 321–334.

Ramírez-Neria, M., González-Sierra, J., Madonski, R., Ramírez-Juárez, R., Hernández-Martínez, E., & Fernández-Anaya, G. (2023). Leader-follower formation and disturbance rejection control for omnidirectional mobile robots. *Robotics*, *12*(5), 102.

Sun, Y., Gao, C., Wu, L., & Yang, Y. (2023). Fuzzy observer-based command filtered tracking control for uncertain strict-feedback nonlinear systems with sensor faults and event-triggered technology. *Nonlinear Dynamics*, *111*(6), 5305–5323.

Vo, A. T., Truong, T. N., & Kang, H.-J. (2022). A novel prescribed-performance-tracking control system with finite-time convergence stability for uncertain robotic manipulators. *Sensors*, *22*(7), 2525.

Yapp, K. K. C., Hoo, C. L., & Lai, C. H. (2024). New anti-windup proportional–integral–derivative for motor speed control. *Asian Journal of Control*, *26*(4), 2123–2138.

Zakeri, E., & Xie, W.-F. (2024). Robust photogrammetry sensor-based real-time pose control of industrial robots. *IEEE/ASME Transactions on Mechatronics*, *29*(5), 3456–3467.

Zhai, D., & Xia, Y. (2022). Initial configurations-independent predefined performance control of robotic manipulator with input saturation. *International Journal of Robust and Nonlinear Control*, *32*(4), 2075–2094.

Zhang, S., Chen, Z., & Zhang, X. (2025). Robust adaptive nonlinear admittance control of robotic manipulators with extra–intra uncertainties and disturbances. *International Journal of Robust and Nonlinear Control*, *35*(4), 2345–2368.

Part IV

Perception and Autonomy

9 Vision and Sensing Perception in Motion

9.1 INTRODUCTION

Robot perception allows machines to sense and comprehend their environment. Robots use perception systems to know their position and nearby obstacles. It helps robots make informed decisions and interact with their environment (Raj & Kos, 2025). While the robot controller decides what actions to execute, the perception system provides situational awareness needed for the robot to perform those actions by observing and analysing the environment (Han et al., 2025). Robots use perception systems to navigate new environments, manipulate objects, understand human commands, and operate safely among humans.

Robots sense their environment using sensor measurements. This could be in the form of RGB images from an onboard camera, range measurements with LiDAR, velocity data from an IMU, etc. Robots use multiple sensors at a time, which together form a sensor suite (Wang et al., 2023). Each sensor measures the environment differently, so having several sensors allows the robot to sense the world in various ways and more robustly. Sensor measurements alone do not provide robots with an understanding of their environment. Perception algorithms like localisation, mapping, and object detection help make sense of sensor measurements (Li et al., 2024).

Processes such as computer vision, signal processing, etc. are required to analyse sensor data (Raj & Kos, 2025). Deep learning techniques have enabled robots to perceive their environments with high accuracy in recent years (Abdullah-Al-Noman et al., 2022). Object detection and tracking are popular perception tasks that allow robots to understand their environment and make decisions in the real world.

Perception is crucial for robots operating in the real world. An autonomous car must be able to detect other cars, pedestrians, and traffic signs always, even when it is raining or sunny. Redundant sensors that measure the same thing using different modalities like cameras and LiDAR are used to provide reliable detections. This process is known as multimodal sensor fusion.

9.2 FOUNDATIONS OF ROBOTIC PERCEPTION

Perception in robotics involves acquiring, processing, interpreting, and understanding data from sensors to detect and make sense of the surrounding environment (Rodríguez-Martínez et al., 2025). Perception systems enable robots to understand their environment and take appropriate actions based on that understanding by converting raw sensor data into meaningful information that can be used for decision-making (Zhao et al., 2025). It allows robots and autonomous systems to operate intelligently in complex real-world environments (Zhao et al., 2025). Perception

DOI: 10.1201/9781003758501-13

plays a crucial role in tasks such as navigation, environment mapping, object recognition, human detection, and interaction with dynamic objects (Liu et al., 2023; Zhang et al., 2023).

Different sensors are used for perception tasks, including cameras, LiDAR sensors, and inertial measurement units (IMUs), which allow robots to perceive their environment (Zhang et al., 2023). Cameras capture visual information from the environment, LiDAR sensors measure distances using laser signals, while IMUs measure motion and orientation (Zhang et al., 2023). Robots can use multiple sensors together to overcome the limitations of individual sensors by leveraging their complementary capabilities (Liu et al., 2023).

Sensor data collected by robots is often raw and unprocessed, noisy, incomplete, or difficult to interpret. Computational methods such as filtering, pattern recognition, and feature extraction are used to process sensor data into useful information that can be used for perception tasks (Xu et al., 2023). Sensing refers to the process of collecting data from sensors, while perception is the process of interpreting sensor data to make sense of the environment (Han et al., 2025).

Sensor data is often uncertain due to various factors such as noise, lighting conditions, occlusions, and other environmental factors (Miki et al., 2022). Perception systems can use techniques such as computer vision, sensor fusion, and filtering to reason about their environment and handle uncertainty to make informed decisions (Wang et al., 2023).

9.3 MODELLING PERCEPTION IN ROBOTIC SYSTEMS

Since robots must reason about their surroundings to act accordingly, we can represent perception as a pipeline as seen in today's perception–planning–control pipelines of autonomous robots (Rodríguez-Martínez et al., 2025; Wang et al., 2025). The Perception–Interpretation–Action (PIA) schema is used to help visualise how a robot senses stimuli from its environment and acts through sequential processes of perception, reasoning, and control stages (Han et al., 2025). Figure 9.1 shows the information flow from sensing to acting from today's schemas that resemble modern perception–decision codesign models (Milojevic et al., 2025).

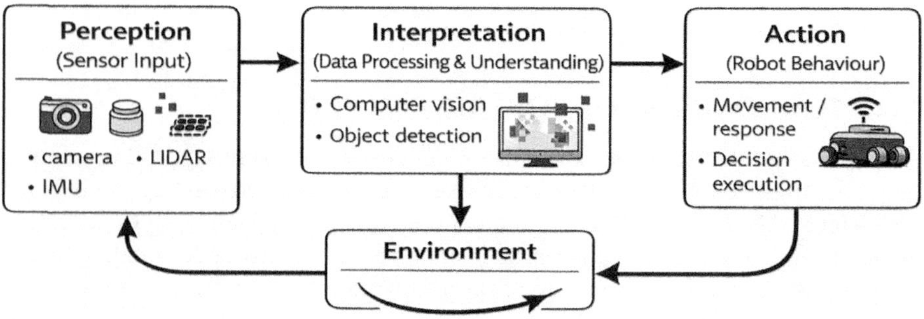

FIGURE 9.1 Perception pipeline in robotics.

Action is initiated by perception (sensing), which makes use of sensors to acquire information about the world as raw data (Zhang et al., 2023). This could come from taking an image with a camera array, measuring distance with LiDAR, or collecting motion information with IMUs. These stages are often paired together to create multimodal perception stacks (Li et al., 2024; Wang et al., 2025). At this stage, we are only gathering information. There is no understanding of what we have collected, like modern-day low-level capture functions of SLAM or dataset pipelines (Li et al., 2024; Zhang et al., 2023).

The next stage is interpretation. During this stage, we process and make sense of the raw information we collected during the perception stage. Tasks such as image processing, feature extraction, and pattern recognition allow us to take in information and gather knowledge from it (Han et al., 2025). Many of these tasks are performed with the use of deep learning. We can take that image we obtained during perception and run it through a 3D detection module to identify objects and boundaries or a semantic mapping module to identify patterns (Zhang et al., 2023). Our system has processed what we sensed about the world and extracted information about it that we can use to make decisions with high-level navigation and manipulation programs (Wang et al., 2025).

Last is action. The system can now act based on its understanding of the world. Since the robot can reason about its surroundings, it can make the decision to turn away from that wall we identified in the image or grab that object of interest with task-driven perception–planning stacks. (Milojevic et al., 2025).

Figure 9.1 depicts it as if sensing would feed computation through perception straight into action; however, Rodríguez-Martínez et al. (2025) model this process as being continuous. Once the robot has acted it will have changed the environment, giving rise to new sensory information which can be used to further update open-loop or closed-loop information. In other words, perception leads to action, which would change the state of the environment we live in as an unceasing loop (Han et al., 2025).

Models like PIA help stress this notion that perception works within a system, allowing for sensing, computation, and control to create action in the environment (Milojevic et al., 2025). Students should now have a greater understanding of how robots can sense information from their environments and turn it into intelligent actions. Perception systems can now be thought of as coupling multimodal sensing with learned decision policies (Han et al., 2025; Wang et al., 2025).

9.4 CASE STUDY: TESLA FULL SELF-DRIVING (FSD)

Tesla Full Self-Driving (FSD) is an example of a perception-driven decision framework that implements computer vision technology. Cameras sense the surroundings, and deep learning models perceive through vision. Tesla FSD then takes control of acceleration, braking, steering, and route navigation (Liu et al., 2022). The machine learning models identify objects like cars, pedestrians, traffic signs, lane markings, etc. (Pavel et al., 2022).

Tesla cars typically have cameras covering 360 degrees around the vehicle. Tesla chose a vision-only policy rather than using lidar/radar along with vision sensors.

The vehicles are equipped with fewer parts, but it depends more on perception using images (Abdullah-Al-Noman et al., 2022).

The problem with this design choice is low-light scenarios, obstructions, and never-seen-before situations can lead to inaccurate perception. Encountering an unknown object may cause it to take the wrong action or no action at all in a timely manner (Abdullah-Al-Noman et al., 2022).

These situations are referred to as perception uncertainty. For example, failing to sense an object or confusing a shadow for an object can result in catastrophic decisions in safety-critical applications (Abdullah-Al-Noman et al., 2022).

Referring to the Perception–Interpretation–Action framework, if there is an issue with perception, that error cascades into the decision and action taken. In this case, the system will not have all the necessary information fed into it to begin with. This limits the functioning of the AI.

9.5 TYPES OF SENSORS IN ROBOTICS

Robotic systems must have one or more sensors to help them understand the world around them. Sensors can provide information about the environment and objects within that environment (Gad et al., 2022). Different sensors allow robots to sense different information about their environment. The application will determine what types of sensors are used. This section will cover common sensors and their advantages and disadvantages.

The camera is one of the most common sensors. Cameras take information in the form of images, which computer vision techniques can process. Robots can detect objects, recognise shapes and characters, and map their environment with the proper programming (Alazeb et al., 2024). Cameras are usually low-cost and are versatile and easy to use in many situations. They often require significant lighting, can be affected by occlusion, and variations in the environment may impact their ability to sense their surroundings (Gad et al., 2022).

LiDAR sensors emit laser signals that bounce back to the sensor. By measuring the time it takes for the signal to return, robots can calculate how far away objects are. Robots can use this information to create robust maps of their environment in three dimensions. LiDAR sensors can be costly, and sometimes weather, such as heavy rain, can impact their abilities (Dreissig et al., 2023).

IMUs consist of accelerometers and gyroscopes to sense motion. Robots can use them to measure velocity, acceleration, and tilt. They are commonly used in drones and self-balancing robots to help track motion. However, without any outside reference, these sensors tend to drift while in use (Yan et al., 2022).

Many robots use ultrasonic and infrared sensors to measure distance. Ultrasonic sensors help robots avoid collisions. Robots can use many sensors simultaneously to help them understand their environments better. This process is known as sensor fusion.

9.6 COMPUTER VISION FUNDAMENTALS

Computer vision refers to the capability of a robotic system to understand what it sees by capturing meaningful information from images or video feeds (Manakitsa et al., 2024). Robots leverage computer vision techniques for navigation, object manipulation, scene understanding, and interacting with humans or moving through complex environments (Wang et al., 2025). Computer vision allows robots to sense their surroundings through visual input and go beyond sensing to higher-level perception (Han et al., 2025).

Detection in computer vision refers to recognising specific objects present in digital images and videos (Manakitsa et al., 2024). Object detection enables robots to detect various objects like obstacles, tools, and humans within their environment in real time (Wang et al., 2025). Detection algorithms allow a robot to locate objects within an image or video frame and draw boundaries around the detected object (Alazeb et al., 2024). This technique uses bounding boxes or segmentation masks to separate the object from the background.

Classification refers to labelling detected objects into specific categories (Manakitsa et al., 2024). For instance, the robot may categorise every identified object as either a car, pedestrian, or traffic sign. Classification allows contextual understanding, enabling robots to make informed decisions based on their observations and assigned tasks (Wang et al., 2025).

Object tracking also relates to detection and refers to the ability to monitor an object's movement from one point to another. Tracking allows a robot to recognise objects of interest and record their movement within its environment (Sahu et al., 2025). Robots can use this technique to predict movement and avoid collisions with moving objects.

Feature extraction describes the process of extracting information from specific parts of an image. These features can include corners, edges, and other unique textures or shapes (Li et al., 2024). Feature extraction allows the system to detect patterns within its environment.

Many different programming libraries, codes, and algorithms are used to identify objects within images and video feeds (Manakitsa et al., 2024). Machine learning models are trained with visual datasets to allow for real-time computing and high-level vision (Han et al., 2025; Wang et al., 2025). With the basics out of the way, we can explore some common image processing techniques.

9.7 IMAGE PROCESSING TECHNIQUES

Image processing techniques help extract useful information from an image and convert it into a format that can be used by a robot to make decisions (Manakitsa et al., 2024). The techniques increase characteristics that help computers detect objects or patterns while filtering out unnecessary information. These manipulations help robots see, identify patterns, extract data, and make decisions. Image processing is vital in computer vision.

Colour detection refers to isolating colours from an image (Abdullah-Al-Noman et al., 2022). It is used when tracking or following specific-coloured lines or paths. Robots can follow coloured lines when the detection system isolates the pixels that make up that colour range (Mohammed et al., 2023).

Edge detection entails determining line endings or points where colours change. This technique uses changes in pixel intensities to identify points where edges occur (Huang & Huang, 2025). Boundaries are mostly the object's external environment, so this technique can help identify objects in images. Edge detection techniques help minimise the amount of data required to represent an image while preserving features that are important for understanding its content.

Contours can be described as continuous curves that follow the edges of an object within an image (Huang & Huang, 2025). Detecting contours allows the system to know the shape and dimensions of the object in view. Systems use this information for several purposes, such as identifying the object or tracking its movement.

Feature extraction techniques try to identify unique features that depict or can describe an image (Li et al., 2024). Features such as corners can help identify similar objects in different images or from frame to frame in a video. Feature-based tracking methods have been used in various applications, such as robot navigation and map-building robots.

Image processing techniques can be used in conjunction with each other. Colour filtering techniques can be used to create a region of interest in the image while edge- and contour-based techniques define the object of interest (Mohammed et al., 2023). Coupling image processing techniques provides better systems capable of perceiving their environment (Alazeb et al., 2024).

Image processing allows us to provide robots with a sense of sight by extracting important information from their environment and representing it in a format that they can understand and act upon (Manakitsa et al., 2024). Robots can navigate and map their environment, track moving objects, and follow lines using these techniques (Han et al., 2025.

9.8 SENSOR FUSION

Sensor fusion can be described as the integration of data from two or more sensors to obtain an improved understanding of an environment (Han et al., 2025). It is true that each sensor provides different levels of precision, and some sensors cannot provide sufficient information about their environment (Xu et al., 2023). With sensor fusion, we can decrease the uncertainty that may come from sensors individually and improve the performance of the robotic system overall (Li et al., 2023).

In robotics, different sensors can provide different fields of information. Cameras are optimal for giving detailed images, but they can perform poorly in dark environments (Rodríguez-Martínez et al., 2025). LiDAR are better at showing distance and measurements, but they don't show as much detail as cameras do (Xu et al., 2023). IMUs help show motion within the environment but can drift too much to be reliable on their own (Huang et al., 2025). Combining all these heterogeneous sensors helps us to complement each other's information and complete the missing pieces (Xu et al., 2023).

Sensor fusion starts by combining all this data together. But first, we must take the time to coordinate and transform all our data to one reference, such as time or coordinate system. This can mean synchronising sensors to measure at the same time or transforming data to the same coordinate frame as other sensors. Other times data will also need to go through filtering techniques to be combined with other sensor data. Ultimately, sensor fusion allows us to create a better representation of the environment than what a single sensor could achieve (Han et al., 2025; Li et al., 2024).

Techniques like filtering allow us to estimate true signals from the noise created by our sensors (Huang et al., 2025). Sensor fusion allows us to perform this technique when there are large amounts of data being collected at different rates.

Autonomous vehicles are a perfect example of sensor fusion. Self-driving cars need to be able to perceive and understand their highly dynamic environments (Han et al., 2025; Li et al., 2023). Cameras can provide the vehicles with RGB details about their environment, LiDAR can provide spatial details about their surroundings, and IMUs can provide information on how the vehicle is moving within the environment.

Robotic perception can never rely on one sensor alone and should always use sensor fusion when applicable (Rodríguez-Martínez et al., 2025). By incorporating many different sources of information, robotic systems can perform better in real-world scenarios (Han et al., 2025).

9.9 PRACTICAL INSIGHT: IMAGE RECOGNITION WITH OpenCV

Computational vision capabilities allow us to see how robots can understand what they see by interpreting what is in front of them in real-time situations. OpenCV is a common computer vision library that teaches us how to implement image processing and object detection with Python code (Sinha et al., 2025). We will be learning how raw data collected from an image can be interpreted using easy examples.

To simplify image recognition, we take in the image by reading it, process the image to understand areas of interest, and applying that object we want the robot to detect such as colour or edge detection. One way a robot could identify an object is through colour detection. The robot could learn what an object looks like by processing the colours within the object. Edge detection allows us to detect specific boundaries (Abdullah-Al-Noman et al., 2022). Edge detection can be applied if we want the robot to recognise the edges of an object.

Below is a simple example of Python code that teaches us the basics of processing an image with the use of the OpenCV library.

Listing 9.1: Basic image processing using OpenCV

```
import cv2

# Load image
image = cv2.imread('image.jpg')

# Convert to grayscale
gray = cv2.cvtColor(image, cv2.COLOR_BGR2GRAY)

# Apply edge detection
edges = cv2.Canny(gray, 100, 200)

# Display results
cv2.imshow('Original', image)
cv2.imshow('Edges', edges)
cv2.waitKey(0)
cv2.destroyAllWindows()
```

In this simple example, we first convert the image to grayscale. We then run edge detection on the image. These edges could potentially be used to detect objects or buildings around you. Additional processing steps could allow you to detect/tracking certain objects using machine learning models to identify what those objects are. OpenCV can also do real-time video analysis. This means your robot could process what it sees in real time. By giving students hands-on examples and allowing them to tinker with processing images, you can show them how concepts are used in real world applications.

9.10 CHALLENGES IN ROBOTIC PERCEPTION

Uncertain information relates to the information acquired from the environment by sensors onboard robots. This data often lacks completeness, truthfulness, and timeliness due to changing lighting, weather, or unforeseen circumstances (Zhang et al., 2023). Such data is often uncertain since it is unlike the structured information robots train on in laboratory conditions (Pandey & Nair, 2025). Real-world information can be noisy and change dynamically, presenting problems with how robots perceive their environment.

Ambiguity in information can occur due to changes in a robot's environment. This includes alterations in lighting conditions, nearby objects, or weather conditions (Dreissig et al., 2023). For example, cameras may be impaired by dark lighting conditions or glare, while lidar may malfunction in fog or heavy rain. Environmental uncertainty can degrade how sensors perceive the environment and make misinterpretations of sensor data more likely.

Noise is another source of uncertainty in perception. Noise can come from inherent imprecision in sensors and is generally defined as random variation in sensor measurements (Dreissig et al., 2023). Because noise is random, it can be challenging to differentiate information that contains noise versus information that does not. Filtering is one commonly used method to remove noise from information.

Occasionally, objects may be blocked from sensors by other objects. This is known as occlusion and can lead to an object not being detected by a robot's perception systems. An example of this would be another vehicle hiding a pedestrian from detection (Ye et al., 2023).

Engineers can implement strategies to prevent failures in robot perception. Sensor fusion, sensor noise filtering, and robust machine learning models can help make perception more accurate (Neves et al., 2023).

9.11 ETHICAL AND REAL-WORLD CONSIDERATIONS

Bias, safety and reliability, privacy, and accountability are ethical and societal considerations of robotic perception systems.

Bias may occur in perception models that are trained on datasets that do not accurately reflect the real world (Raptis et al., 2025). For example, if a model is trained on datasets that do not include certain environments, objects, or people, it may not perform as well when encountered with these situations, leading to biased decisions (Li et al., 2023).

Safety and reliability concerns include the potential for mistakes made by robots due to incorrect perception. For example, if a robot misclassifies an object or fails to detect it, it may perform the wrong action. This could be extremely dangerous if the robot is performing safety-critical tasks, such as driving a car. Ensuring that perception systems work reliably in all situations is extremely important to safety (Kuznietsov et al., 2024).

Privacy concerns stem from sensors that collect data about the world and their surroundings. Cameras can easily be used to collect personal data if not used properly or placed in public areas (Söderlund, 2023).

Accountability is difficult for machine learning-based systems. The people who design systems must be accountable for ensuring their robots are safe, do not discriminate against people, and perform reliably to avoid injuring people (Tóth et al., 2022).

9.12 CONCLUSION

This chapter provided insight into how vision and sensing allow robots to see and understand their environment. Vision, LiDAR and IMUs were examples of sensors enabling perception by providing necessary data. Cameras, LiDAR, and computer vision algorithms were presented to help process the data into usable information. The Perception–Interpretation–Action model was introduced to understand how a robot takes in sensory data and turns it into actions. Tesla FSD serves as an example that vision enables great capabilities but can fail or fall short when met with certain conditions. Object detection/tracking, image processing, and sensor fusion were discussed as necessary steps to allow a robot to have a correct perception of its environment and reduce uncertainty. Noise, bias, and changing environments were highlighted as challenges with perceiving the real world. To conclude, vision and sensing allow robots to understand their environment through many different sensors and processing techniques. Proper design is necessary to allow for accurate, reliable, and safe perception.

9.13 DISCUSSION QUESTIONS

1. How do robots sense their surroundings?
2. What are the drawbacks of camera-based perception systems?
3. Why is sensor fusion necessary for perception?
4. In what ways can bias impact perception systems?
5. What are some ethical concerns regarding vision systems?

9.14 EXERCISES

Exercise 1: Image Processing

Manipulate an image using OpenCV library by applying edge detection or colour filtering. Discuss your findings.

Exercise 2: Compare Sensors

Research and compare the accuracy, price, and limitations of various sensors. Some sensors to consider are camera, LiDAR, and IMU.

Exercise 3: Perception Failures

Find an example of when a robot did not perceive its surroundings correctly. What could have been the cause? How can it be improved?

Exercise 4: Perception System Writeup

Write about how the weather and bias in data can impact the way robots perceive their surroundings and make decisions.

REFERENCES

Abdullah-Al-Noman, M., Eva, A., Yeahyea, T. B., & Khan, R. (2022). Computer vision-based robotic arm for object color, shape, and size detection. *Journal of Robotics and Control*, *3*(2), 137–149.

Alazeb, A., Chughtai, B. R., Al Mudawi, N., Alqahtani, Y., Alonazi, M., Aljuaid, H., Jalal, A., & Liu, H. (2024). Remote intelligent perception system for multi-object detection. *Frontiers in Neurorobotics*, *18*, 1387105.

Dreissig, M., Scheuble, D., Piewak, F., & Boedecker, J. (2023). Survey on LiDAR perception in adverse weather conditions. In *2023 IEEE Intelligent Vehicles Symposium (IV)* (pp. 1–8). IEEE.

Gad, A., Basmaji, T., Yaghi, M., AlHeeh, H., Alkhedher, M., & Ghazal, M. (2022). Multiple object tracking in robotic applications: Trends and challenges. *Applied Sciences*, *12*(19), 9949.

Han, X., Chen, S., Fu, Z., Feng, Z., Fan, L., An, D., Wang, C., Guo, L., Meng, W., Zhang, X., Xu, R., & Xu, S. (2025). Multimodal fusion and vision-language models: A survey for robot vision. *Information Fusion*, *110*, 102438.

Huang, G., Huang, H., Zhai, Y., Tang, G., Zhang, L., Gao, X., Huang, Y., & Ge, G. (2025). Multi-sensor fusion for wheel-inertial-visual systems using a fuzzification-assisted iterated error state Kalman filter. *Sensors*, *25*(3), 1120.

Huang, Q., & Huang, J. (2025). Comprehensive review of edge and contour detection: From traditional methods to recent advances. *Neural Computing and Applications*, *37*(5), 4123–4152.

Kuznietsov, A., Gyevnar, B., Wang, C., Peters, S., & Albrecht, S. V. (2024). Explainable AI for safe and trustworthy autonomous driving: A systematic review. *IEEE Transactions on Intelligent Transportation Systems*, *25*(6), 5304–5325.

Li, J., Wang, B., Ma, H., Gao, L., & Fu, H. (2024). Visual feature extraction and tracking method based on corner flow detection. *IECE Transactions on Intelligent Systematics*, *17*(2), 101–112.

Li, Q., Yuan, Z., & Huai, J. (2023). Multi-sensor fusion for robust localization with moving object segmentation in complex dynamic 3D scenes. *International Journal of Applied Earth Observation and Geoinformation*, *124*, 103540.

Liu, F., Lu, Z., & Lin, X. (2022). Vision-based environmental perception for autonomous driving. *Proceedings of the Institution of Mechanical Engineers, Part D: Journal of Automobile Engineering*, *236*(12), 2749–2770.

Liu, Y., Fu, Y., Qin, M., Xu, Y., Xu, B., Chen, F., Goossens, B., Sun, P. Z., Yu, H., Liu, C., Chen, L., Tao, W., & Zhao, H. (2023). BotanicGarden: A high-quality dataset for robot navigation in unstructured natural environments. *IEEE Robotics and Automation Letters*, *8*(7), 3925–3932.

Manakitsa, N., Maraslidis, G. S., Moysis, L., & Fragulis, G. (2024). A review of machine learning and deep learning for object detection, semantic segmentation, and human action recognition in machine and robotic vision. *Technologies*, *12*(2), 15.

Miki, T., Hwangbo, J., Wellhausen, L., Koltun, V., & Hutter, M. (2022). Learning robust perceptive locomotion for quadrupedal robots in the wild. *Science Robotics*, *7*(62), eabk2822.

Milojevic, D., Zardini, G., Elser, M., Censi, A., & Frazzoli, E. (2025). CODEI: Resource-efficient task-driven codesign of perception and decision making for mobile robots applied to autonomous vehicles. *IEEE Transactions on Robotics*, *41*(4), 2899–2919.

Mohammed, M., AL-Khafaji, M., & Abbas, T. (2023). Smart robot vision for a pick and place robotic system. *Engineering and Technology Journal*, *41*(1), 34–47.

Neves, F., Claro, R., & Pinto, A. (2023). End-to-end detection of a landing platform for off-shore UAVs based on a multimodal early fusion approach. *Sensors*, *23*(5), 2540.

Pandey, S., & Nair, S. B. (2025). RoboDA: A dataset for domain adaptation in robot vision. *IETE Technical Review*, *42*(3), 456–470.

Pavel, M. I., Tan, S. Y., & Abdullah, A. (2022). Vision-based autonomous vehicle systems based on deep learning: A systematic literature review. *Applied Sciences*, *12*(14), 6831.

Raj, R., & Kos, A. (2025). An extensive study of convolutional neural networks: Applications in computer vision for improved robotics perceptions. *Sensors*, *25*(2), 1234.

Raptis, E. K., Kapoutsis, A. C., & Kosmatopoulos, E. B. (2025). Agentic LLM-based robotic systems for real-world applications: A review on their agenticness and ethics. *Frontiers in Robotics and AI*, *12*, 1538240.

Rodríguez-Martínez, E. A., Flores-Fuentes, W., Achakir, F., Sergiyenko, O., & Murrieta-Rico, F. N. (2025). Vision-based navigation and perception for autonomous robots: Sensors, SLAM, control strategies, and cross-domain applications—A review. *Eng*, *6*(7), 153.

Sahu, U. K., Mebin, K. S., Abhinav, K., Muzammil, M. P., Jaiswal, A., Yadav, U. K., & Dash, S. K. (2025). Autonomous object tracking with vision-based control using a 2DOF robotic arm. *Scientific Reports*, *15*, 4562.

Sinha, E., Kumar, A., & Tyagi, A. (2025). OpenCV for computer vision applications. *International Journal for Multidisciplinary Research*, *7*(5), 101–110.

Söderlund, M. (2023). Service robots and artificial morality: An examination of robot behavior that violates human privacy. *Journal of Service Theory and Practice*, *33*(5), 569–590.

Tóth, Z., Caruana, R., Gruber, T., & Loebbecke, C. (2022). The dawn of the AI robots: Towards a new framework of AI robot accountability. *Journal of Business Ethics*, *179*(1), 207–229.

Wang, S., Nikolić, M., Lam, T., Gao, Q., Ding, R., & Zhang, T. (2025). Robot manipulation based on embodied visual perception: A survey. *CAAI Transactions on Intelligence Technology*, *10*(3), 845–872.

Wang, Z., Liu, Y., Duan, Y., Li, X., Zhang, X., Ji, J., Dong, E., & Zhang, Y. (2023). USTC FLICAR: A sensor fusion dataset of LiDAR-inertial-camera for heavy-duty autonomous aerial work robots. *The International Journal of Robotics Research*, *42*(5), 987–1005.

Xu, J., Li, B., Yang, S., Zhi, Y., & Li, R. (2023). SLI-SLAM: Autonomous navigation and accurate mapping for quadruped robots in complex environments using LiDAR, stereo camera, and IMU fusion. In *2023 Chinese Control Conference (CCC)* (pp. 4567–4574). IEEE.

Yan, Y., Zhang, B., Zhou, J., Zhang, Y., & Liu, X. (2022). Real-time localization and mapping utilizing multi-sensor fusion and visual–IMU–wheel odometry for agricultural robots in unstructured, dynamic and GPS-denied greenhouse environments. *Agronomy, 12*(8), 1740.

Ye, Y., Ma, X., Zhou, X., Bao, G., Wan, W., & Cai, S. (2023). Dynamic and real-time object detection based on deep learning for home service robots. *Sensors, 23*(23), 9752.

Zhang, A., Eranki, C., Zhang, C., Park, J.-H., Hong, R., Kalyani, P., Kalyanaraman, L., Gamare, A., Bagad, A., Esteva, M., & Biswas, J. (2023). Toward robust robot 3-D perception in urban environments: The UT Campus Object Dataset. *IEEE Transactions on Robotics, 39*(6), 4470–4489.

Zhao, W., Gangaraju, K., & Yuan, F. (2025). Multimodal perception-driven decision-making for human–robot interaction: A survey. *Frontiers in Robotics and AI, 12*, 1543210.

10 Navigation and Mapping the Unknown

10.1 INTRODUCTION

Navigation enables robots to move to where they want to go by themselves (Yan et al., 2022). Navigation is different from simply moving around following instructions. Navigation requires localisation: knowing where the robot is. Navigation also requires that the robot understand its environment and plan how to reach its goal (Qiao et al., 2024). Robots must navigate autonomously in unknown environments, which may change over time (Zhang et al., 2023).

Navigation requires solving two critical problems, localisation and map building (Qiao et al., 2024). Localisation is the ability to know where a robot is, while mapping is the ability to know about its environment (Zhou et al., 2023). In general, robots must solve both problems at the same time because the robot does not know its environment in advance.

Navigation extends perception and control. Recall from previous chapters that perception allows a robot to sense its environment, and control allows a robot to move around. Navigation uses the output of the perception system to move around in the environment. Navigation also requires the robot to decide where to move and how to get there (Chen et al., 2024).

When robots navigate in the real world, there are many complications. For instance, there may be noise in the sensor data, the world may change, and there may be limited resources for computation (Liu et al., 2022). For example, if a robot is exploring the rooms of a building, it may find that furniture has been moved since the map was created. Additionally, the robot may not know its location precisely because of noise in the sensor data (Chen et al., 2024).

This chapter presents an introduction to navigation and map building. We will see how robots create maps of their environment that they can use to operate in unknown spaces (Qiao et al., 2024). This chapter covers SLAM, odometry, and path planning (Yan et al., 2022).

10.2 FOUNDATIONS OF ROBOTIC NAVIGATION

Navigation is the process of moving a system from the initial pose to goal pose while avoiding obstacles or collisions (Jiang et al., 2022). Navigation implies that your system can sense where it is now, know where it wants to go and actuate itself to reach its destination (Sun et al., 2023). Robots navigate environments that are known or unknown.

Localisation describes where a robot is in its environment (Qiao et al., 2024; Zhao et al., 2022), and mapping describes the environment around the robot (Sousa et al., 2023). Navigation breakdown in known environments consist of localisation

DOI: 10.1201/9781003758501-14

and path planning since robots are given a map (Sun et al., 2023). When navigating unknown environments, robots must create their own map, which is called simultaneous localisation and mapping or SLAM (Jiang et al., 2022; Zhao et al., 2022).

Maps may be constructed as grid map or feature map (Liu et al., 2023). Path planning defines how your robot goes from point A to point B either offline or in real time (Wang et al., 2025). These are what allow robots to safely navigate the world around them intelligently (Huang & Geng, 2024).

10.3 MODELLING NAVIGATION WITH SLAM

Robots cannot autonomously navigate without knowing the environment. Mapping and localisation enable robots to navigate unknown environments (Chen et al., 2024). The SLAM loop model illustrates the mapping and localisation processes (Rodríguez-Martínez et al., 2025). Localisation is the process of estimating the robot's pose based on the map. Mapping involves finding the locations of features in the environment while localising. The mapping and localisation process involve several steps. The first step is observation, which occurs when the robot senses the environment using onboard sensors (Zhu et al., 2023). Sensors may detect information about the distance from other objects, such as LiDAR and capture images using cameras. In motion-based mapping and localisation techniques, encoders and inertial measurement units (IMUs) can also be used to make observations about the robot's movement. Observations can also show changes to the environment. Map update is the second step in the mapping and localisation process (Mao et al., 2025). When the robot makes observations about the environment, it can either build a map or update an existing map (Zhou et al., 2023). Mapping helps the robot know the locations of features in the environment such as obstacles and walls (Mao et al., 2025). The map is updated continuously as the robot makes more observations using its sensors. Finally, the robot relocates itself within the map in the re-localisation step (Chen et al., 2024). Robots perform localisation to estimate their pose based on the map. Accurate localisation allows robots to understand their environments properly (Yang et al., 2026).

Figure 10.1 Abstracts this iterative process into a single loop: observe → update map → localise → gain knowledge of the environment, which will affect future observations. Through this loop, localisation and mapping can continuously improve. It should be clear from this loop that SLAM tightly couples mapping and localisation. In fact, poorly localised poses will create inconsistencies in map generation, so it is important to keep track of our uncertainty as well as maintaining consistency. The robot can achieve this by continuing to incrementally observe its environment. This model can start to think about how robots can map their environments as they navigate through them. This serves as groundwork for the localisation and mapping methods discussed later.

10.4 CASE STUDY: TurtleBot NAVIGATION USING ROBOT OPERATING SYSTEM

The TurtleBot is a popular open-source mobile robot platform that is often used for research and education on mobile robotics and autonomous navigation tasks (Ghazal et al., 2024). Since TurtleBot runs ROS (Robot Operating System), we can also use TurtleBot to exemplify how SLAM/navigation algorithms implemented in ROS

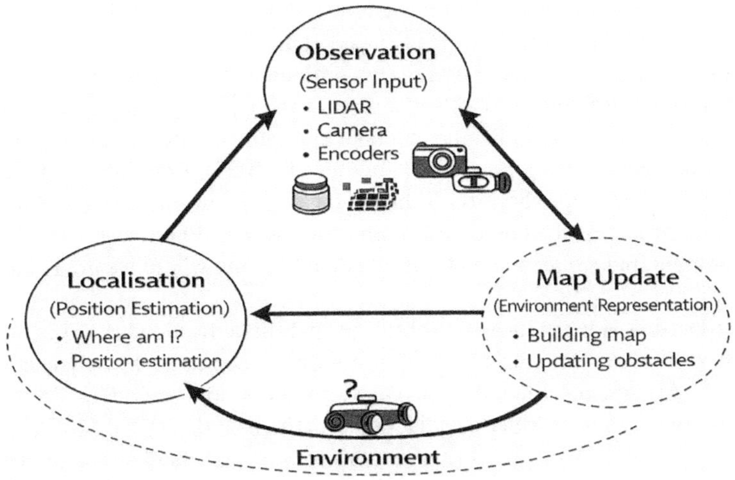

FIGURE 10.1 SLAM process in autonomous robotics.

operate on a real-world system (Zhao et al., 2022). TurtleBot is simulated here using Gazebo environment for example purpose (Ghazal et al., 2024).

Assuming we want to explore an unknown indoor environment and map it using TurtleBot, initially the robot has no prior knowledge of the map/environment. Using onboard sensors such as LiDAR and encoders, we move around the environment to explore it (Sanboontho et al., 2025). Sensors send information to SLAM algorithm to construct the map and localise itself (Sun et al., 2023).

While exploring the environment with each movement, the robot keeps mapping the unknown environment (build a map with walls, obstacles, and free spaces) and localising itself with respect to the map at the same time (Sanboontho et al., 2025). Localisation can be done with various localisation algorithms (fingerprinting, AMCL, etc.) (Li et al., 2023).

Notice that ROS makes communication possible between different nodes. Here, individual nodes can be assigned to handle information from sensors (perception node), SLAM algorithm (mapping and localisation), and planning and control (navigate node) (Zhao et al., 2022). Thanks to ROS, information can be communicated in real-time between these nodes.

After mapping the environment and localising itself with respect to the map, the robot should plan a path to reach some target location while avoiding obstacles/known walls (Li et al., 2024). Refer to the Path Planning code/video for an example of how path planning can be done using TurtleBot. If we made some changes to the environment (i.e. added boxes to the scene), we can re-plan the path as shown in the Replan Path video.

10.5 SLAM FUNDAMENTALS

SLAM refers to a set of techniques used in robotics which allows a robot to construct or build a map of an unknown environment while at the same time keeping track of its own location within that map (Qiao et al., 2024; Zheng et al., 2023). SLAM systems allow robots to navigate through an unknown environment using no prior information about their surroundings (Placed et al., 2022).

Essentially, SLAM involves solving two problems at the same time. Firstly, mapping refers to the problem of building a map of the environment based on observations gathered by sensors. This map can take many forms and can consist of features like walls, obstacles, and landmarks (Zhou et al., 2023). Localisation refers to the problem of determining where the robot is located based on its map (Zheng et al., 2023). These problems depend on each other; we need to know where we are to map an environment, but we also need to have a map to know where we are (Placed et al., 2022).

Typically, SLAM is performed within a probabilistic framework. This means that uncertainty in sensor measurements and motion is accounted for when implementing SLAM. All sensors, including LiDAR, cameras, and encoders, contain some amount of noise or uncertainty (Rodríguez-Martínez et al., 2025). Motion also does not happen exactly as we want it to due to noise, such as wheel slippage or irregular terrain. SLAM tackles this uncertainty by using probability and statistics to estimate the likely map and location (Qiao et al., 2024; Zheng et al., 2023).

Data association is an important concept in SLAM. When a robot observes a feature in the environment, it must be able to associate that feature with what is stored in its map. If a robot observes a landmark or wall, it needs to know whether it has encountered that feature before (Wu et al., 2023). Proper data association allows the SLAM system to maintain a consistent map (Wu et al., 2023).

Loop closure is another important SLAM concept. Loop closure refers to the ability to detect when a robot returns to a location that it has previously mapped (Tsintotas et al., 2022). By identifying this, we can reduce the uncertainty in both the map and the robot's location (Zhou et al., 2023). Without loop closures, small mapping and localisation errors will grow larger over time (Tsintotas et al., 2022).

SLAM allows for robots and robots to autonomously navigate through unknown environments (Rodríguez-Martínez et al., 2025; Zheng et al., 2023). By simultaneously computing the map of the environment and their location on that map, SLAM forms the basis of robotic navigation techniques (Placed et al., 2022).

10.6 ODOMETRY AND DRIFT CORRECTION

Odometry refers to a robot's ability to understand where it is by using the data from its own movements (Tran & Ryoo, 2025). This usually involves using wheel encoders to measure wheel rotation and using that information to estimate the robot's travelled distance and change in orientation (Huang et al., 2024). Odometry allows the robot to keep track of how far and where it has moved from an initial position in a simple and computationally inexpensive way.

Since odometry is only based on the robot's own movement data, errors in the robot's position will compound over time, which is known as drift (Tran & Ryoo, 2025). This can occur for a variety of reasons such as errors in sensor data, wheels slipping, driving on uneven ground, or manufacturing defects. Unless the robot's estimated position is corrected, the drift will become larger over time and cause the robot to think it is in the wrong location (Yan et al., 2022).

Drift can be corrected by using other sensor data such as LiDAR or camera observations to better estimate the robot's pose (Huang et al., 2024). Odometry can give an estimate of the robot's pose between sensor observations, and when the next observation is received the estimate can be compared to the observation to compute how much drift has occurred. The map and robot's pose can then be corrected.

When a robot traverses its environment, it may traverse the same area twice. Loop closure refers to when the robot recognises that it is in a location that it has previously been (Zhang et al., 2024). Loop closures can help reduce errors in mapping and localisation by providing constraints that can be used to correct the map and the robot's estimated pose (Wang et al., 2025).

Multiple sensors can be combined to get better estimates of the robot's pose; this is known as sensor fusion (Yan et al., 2022). Commonly used sensors include encoders, IMUs, and LiDAR. When these sensors are used together, each sensor can provide unique information that can help correct errors in the data from other sensors (Tran & Ryoo, 2025).

10.7 SENSOR FUSION IN NAVIGATION

Sensor fusion is required for autonomous robot navigation due to no single sensor being able to perceive its environment completely and accurately. Fusion techniques use a variety of sensors to overcome different sensors' weaknesses (Zhu et al., 2023).

Wheel encoders can calculate a robot's approximate motion by odometry but will drift as time progresses (Yan et al., 2022). Distance sensors such as LiDAR can give highly accurate distance measurements of surrounding points but lack information about features, while cameras provide dense features about objects and landmarks in the environment and IMUs sense changes in movement (Huang et al., 2025).

Applications of sensor fusion can include mapping and localisation. Odometry data can be fused with LiDAR or camera data to help compensate for drift in position and pose estimation. GPS data can also be fused with accelerometer data to localise a robot more accurately (Tran & Ryoo, 2025). Essentially, sensor fusion allows you to take two different uncertainties and combine them to lower your overall uncertainty.

Measurements from different sensors need to be brought into the same time and reference frame before being passed into a fusion method. This can be done by synchronising your measurements and transforming them into the same coordinate frame (Li et al., 2024; Zhu et al., 2023).

Sensor fusion allows robots to operate reliably in unknown environments or when individual sensors may be noisy or inaccurate (Liu et al., 2023; Tran & Ryoo, 2025; Yan et al., 2022).

10.8 PATH PLANNING AND OBSTACLE AVOIDANCE

Path planning is how a robot plans on moving from its location to some goal without hitting any obstacles in the way (Rafai et al., 2022). It allows robots to move freely and is necessary for robot navigation. Path planning works closely with localisation and perception systems to function (Jiang, 2025; Wei et al., 2025).

Maps, localisation and sensor data are used together to plan paths. If robots know their environment, optimal paths to goals can be planned to use maps to reduce travel distance/time/memory (Hentout et al., 2024). These paths can be changed if the map becomes incorrect or if the robot becomes aware of its environment (Benmachiche et al., 2025).

Global planning and local planning can be thought of as subcomponents of path planning. Global planning creates a path for the robot to some goal from its position using a map of the environment. Local planning, however, only accounts for where the robot is currently at and uses sensors to dictate how the robot should move (Sun et al., 2023; Yang et al., 2023).

Avoiding obstacles is necessary for any robot moving through an environment. Robots will use sensors to tell if there is an obstacle in their way and change paths to not run into that object. Depending on how fast the robot detects something in its path, it may slow down or change directions to avoid it (Wang et al., 2025).

Path planning is typically not a one-time occurrence. As the robot moves, it may receive new information that will change the model of its environment (Benmachiche et al., 2025; Jiang, 2025).

Path planning and obstacle avoidance go hand in hand to allow robots to travel from point A to point B (Yang et al., 2023)

10.9 PRACTICAL INSIGHT: ROS-BASED NAVIGATION SIMULATION

Simulating a robot exploring a space also helps students visualise how mapping, localisation, and path planning functions work together (Ghazal et al., 2024). By using TurtleBot on the Robot Operating System along with Gazebo simulation software, students can watch the robot explore and navigate through a space in real-time.

Students can set the robot's starting location within a space that it has not mapped (Sun et al., 2023). Once the simulation begins, LiDAR sensors provide readings as the robot navigates the space. SLAM then uses that data to localise itself and map out the space (Sun et al., 2023; Zhao et al., 2022). Mapping the space becomes more complete as the robot continues to move to new areas (Ghazal et al., 2024).

After mapping the space, students can input coordinates to navigate to that location. Path planning algorithms go into effect to plot the most optimal path to that location (Li et al., 2024). The robot will localise itself as it travels to that destination and will take action to move around obstacles or other changes to its environment (Sun et al., 2023).

An example of what a very simple navigation program's inputs may look like in Python:

Listing 10.1: Conceptual ROS navigation workflow

```
# Pseudo-code for navigation process

initialize_robot()
start_sensors()
```

```
while robot_is_active:
    sensor_data = get_sensor_data()
    map = update_map(sensor_data)
    position = estimate_position(map, sensor_data)

    if goal_set:
        path = plan_path(position, goal, map)
        move_robot_along(path)
```

The above example shows sensing → mapping → localisation → motion planning looping forever. ROS manages each process as separate nodes talking to each other in real time.

Simulation enables students to play with examples and see how system responds to parameter tuning or environmental changes for navigation tasks. One of the biggest advantages of simulation is that it allows you to try things out without having to build robot hardware. Students will learn how autonomous robots navigate the world from these examples.

10.10 CHALLENGES IN ROBOTIC NAVIGATION

Robotic navigation in real-world environments can encounter several problems due to uncertainty, environmental conditions, and system limitations (Rodríguez-Martínez et al., 2025; Wang et al., 2025). While SLAM and sensor fusion can provide robust solutions to these problems, it's important to take these challenges into consideration when navigating in the real world.

Sensor noise and data uncertainty is one source of problems that can occur during robotic navigation. Many sensors, such as LiDAR, cameras, and encoders, can provide noisy or inaccurate data due to miscalibration, environmental conditions, or other limitations of the sensor itself (Yao et al., 2024). This noise and uncertainty can lead to inaccuracy during both mapping and localisation.

Changes to the environment or map inconsistency can also cause problems during navigation. The real world is dynamic, so changes to the world, such as moving objects like people and vehicles, can quickly change the landscape and cause previous maps to become inaccurate (Zohaib et al., 2023). Robots will also need to detect these changes and adapt to them accordingly (Chen et al., 2024; Zhou et al., 2023).

Localisation error and inconsistent maps can occur during mapping. When a robot misinterprets where it is located on the map, it can cause the map to become warped and inconsistent (Zhang et al., 2024). Loop closure can help reduce localisation error in maps, but ensuring consistency is still an issue (Zhang et al., 2024).

Performance can also be affected by the robots' computational limitations as well as environmental factors such as lighting conditions and terrain (Rodríguez-Martínez et al., 2025; Wang et al., 2025).

10.11 ETHICAL AND REAL-WORLD CONSIDERATIONS

Ethical navigation systems consider how robots navigate safely and reliably through environments they share with humans and other robots (Neupane et al., 2023). This includes ensuring that robots don't collide with people or objects, follow safe paths that won't cause harm, and can be trusted to behave predictably (Tóth et al., 2022).

Safety: Navigation should be safe for both robots and humans in the environment. Robots should be able to sense their surroundings and avoid obstacles, including people, in real time. Localisation, mapping, and planning failures can cause robots to collide with objects or behave unpredictably. Navigation systems should include fail-safe behaviours and be operable only within known safe limits. Navigation systems should be particularly safe in environments like hospitals, warehouses, or public places.

Reliability: Robots should be able to navigate reliably even in uncertain environments. Maps may be wrong, sensor data is often noisy, and perfect paths do not always exist (Neupane et al., 2023). Unpredictable or unreliable robot navigation can decrease user trust and limit deployments in real-world settings (Raptis et al., 2025).

Trust and transparency: Users should trust that a robot will safely and predictably navigate when deployed. Robots should be well tested and able to handle edge cases that it could encounter during operation. In some situations, explaining system decisions may increase trust or user understanding of the system (Schött et al., 2023).

Accountability: If a navigation system causes an accident or fails in some way, who is responsible? Is it the developers, the manufacturers, or the operators? Ensuring accountability for ethical navigation is challenging but important (Neupane et al., 2023). Navigation systems should be designed responsibly and deployed with care by operators (Tóth et al., 2022).

Considering these questions is important not just for building effective navigation systems but also for ensuring they're deployed responsibly. Navigation isn't just a technical challenge; it's a social one as well. By considering safety, reliability, and trust, we can make sure our robots will operate safely and as intended in real-world environments (Raptis et al., 2025).

10.12 CONCLUSION

This chapter focused on exploring robot navigation within unknown environments by combining localisation and decision-making components using perception. The concepts of SLAM were introduced, showing how a robot can construct maps while simultaneously keeping track of its location. Odometry, drift correction, and sensor fusion were also discussed as ways of fusing multiple sources of information to reduce uncertainty. The SLAM loop was presented, outlining how observation, map updates, and localisation work together. An example of using the TurtleBot within ROS was provided. Path planning and obstacle avoidance algorithms were introduced to show how robots can identify safe paths through their environment. Factors such as sensor noise, dynamic environments, and computational limitations were mentioned as real-world challenges. Ethics was also briefly discussed regarding safety and trust. Key takeaways from the chapter would be that navigation is achieved by fusing multiple systems and continuously updating information to allow a robot to operate within uncertain environments.

10.13 DISCUSSION QUESTIONS

1. How can a robot explore its environment if it does not know its initial location or have a map of the environment?
2. How are localisation and mapping related in the context of SLAM?
3. Why can't we rely solely on odometry for navigation?
4. How does sensor fusion help improve navigation?
5. What are some challenges that can arise when navigating in dynamic environments?

10.14 EXERCISES

Exercise 1: SLAM Simulation

Simulate a robot exploring an environment using a ROS-based simulator such as Gazebo. Visualise how the map of the environment is built over time as the robot explores.

Exercise 2: Odometry Drift

Simulate a robot moving using odometry data. Track how the error in the robot's estimated position accumulates over time.

Exercise 3: Path Planning

Create a path for a robot to navigate through a known environment. Consider factors such as obstacles in the environment and the efficiency of the path.

Exercise 4: Navigation Systems Essay

Discuss the various methods that robots can use to create maps of their environment even when they are uncertain of their location. What makes you trust or not trust these systems?

REFERENCES

Benmachiche, A., Derdour, M., Kahil, M. S., Ghanem, M. C., & Deriche, M. (2025). Adaptive hybrid PSO–APF algorithm for advanced path planning in next-generation autonomous robots. *Sensors*, *25*(18), 6212.

Chen, B., Zhong, X., Xie, H., Peng, P., Hu, H., Zhong, X., & Liu, Q. (2024). SLAM-RAMU: 3D LiDAR-IMU lifelong SLAM with relocalization and autonomous map updating for accurate and reliable navigation. *Industrial Robot*, *51*(1), 1–18.

Ghazal, M., Al-Ghadhanfari, M., & Waisi, N. (2024). Simulation of autonomous navigation of TurtleBot robot system based on robot operating system. *Bulletin of Electrical Engineering and Informatics*, *13*(2), 1012–1021.

Hentout, A., Maoudj, A., & Kouider, A. (2024). Shortest path planning and efficient fuzzy logic control of mobile robots in indoor static and dynamic environments. *Romanian Journal of Information Science and Technology, 27*(1), 67–86.

Huang, G., Huang, H., Zhai, Y., Tang, G., Zhang, L., Gao, X., Huang, Y., & Ge, G. (2024). Multi-sensor fusion for wheel-inertial-visual systems using a fuzzification-assisted iterated error state Kalman filter. *Sensors, 24*(23), 8123.

Huang, Z., & Geng, J. (2024). Research on laser-based mobile robot SLAM and autonomous navigation. *Journal of Physics: Conference Series, 2683*, 012031.

Huang, Z., Ye, G., Yang, P., & Yu, W. (2025). Application of multi-sensor fusion localization algorithm based on recurrent neural networks. *Scientific Reports, 15*, 12345.

Jiang, S. (2025). Path planning and navigation technologies for autonomous mobile robots in dynamic indoor environments. *Highlights in Science, Engineering and Technology, 52*, 210–224.

Jiang, S., Wang, S., Yi, Z., Zhang, M., & Lv, X. (2022). Autonomous navigation system of greenhouse mobile robot based on 3D lidar and 2D lidar SLAM. *Frontiers in Plant Science, 13*, 852725.

Li, H., Zou, Y., Chen, N., Lin, J., Liu, X., Xu, W., Zheng, C., Li, R., He, D., Kong, F., Cai, Y., Liu, Z., Zhou, S., Xue, K., & Zhang, F. (2024). MARS-LVIG dataset: A multi-sensor aerial robots SLAM dataset for LiDAR–visual–inertial–GNSS fusion. *The International Journal of Robotics Research, 43*(2), 145–156.

Li, N., Zhou, F., Yao, K., Hu, X., & Wang, R. (2023). Multisensor fusion SLAM research based on improved RBPF-SLAM algorithm. *Journal of Sensors, 2023*(1), 16.

Liu, G. P. C. J., Rezende, A. M. C., Miranda, V. R. F., Fernandes, R., Azpúrua, H., Neto, A. A., Pessin, G., & Freitas, G. (2022). EKF-LOAM: An adaptive fusion of LiDAR SLAM with wheel odometry and inertial data for confined spaces with few geometric features. *IEEE Transactions on Automation Science and Engineering, 19*(3), 2100–2115.

Liu, Y., Fu, Y., Qin, M., Xu, Y., Xu, B., Chen, F., Goossens, B., Sun, P. Z., Yu, H., Liu, C., Chen, L., Tao, W., & Zhao, H. (2023). BotanicGarden: A high-quality dataset for robot navigation in unstructured natural environments. *IEEE Robotics and Automation Letters, 8*(7), 4210–4217.

Mao, W., Jiang, L., Liu, S., Xi, S., & Bao, H. (2025). A long-term localization and mapping system for autonomous inspection robots in large-scale environments using 3D LiDAR sensors. *PLOS ONE, 20*(7), e0312345.

Neupane, S., Mitra, S., Fernandez, I. A., Saha, S., Mittal, S., Chen, J., Pillai, N., & Rahimi, S. (2023). Security considerations in AI-robotics: A survey of current methods, challenges, and opportunities. *IEEE Access, 11*, 108743–108772.

Placed, J. A., Strader, J., Carrillo, H., Atanasov, N. A., Indelman, V., Carlone, L., & Castellanos, J. A. (2022). A survey on active simultaneous localization and mapping: State of the art and new frontiers. *IEEE Transactions on Robotics, 38*(4), 2015–2041.

Qiao, J., Guo, J., & Li, Y. (2024). Simultaneous localization and mapping (SLAM)-based robot localization and navigation algorithm. *Applied Water Science, 14*, 233.

Rafai, A. N. A., Adzhar, N., & Jaini, N. (2022). A review on path planning and obstacle avoidance algorithms for autonomous mobile robots. *Journal of Robotics, 2022*, 1–24.

Raptis, E. K., Kapoutsis, A. C., & Kosmatopoulos, E. B. (2025). Agentic LLM-based robotic systems for real-world applications: A review on their agenticness and ethics. *Frontiers in Robotics and AI, 12*, 1502345.

Rodríguez-Martínez, E. A., Flores-Fuentes, W., Achakir, F., Sergiyenko, O., & Murrieta-Rico, F. N. (2025). Vision-based navigation and perception for autonomous robots: Sensors, SLAM, control strategies, and cross-domain applications—A review. *Eng, 6*(3), 1230–1278.

Sanboontho, W., Pichitkul, A., Tantrairatn, S., Phiboon, T., & Ariyarit, A. (2025). Indoor localization and measurement object in field for TurtleBot using combined 2D-LiDAR and monocular camera. *Journal of Physics: Conference Series, 2875*, 012045.

Schött, S., Amin, R. M., & Butz, A. (2023). A literature survey of how to convey transparency in co-located human-robot interaction. *Multimodal Technologies and Interaction*, *7*(2), 19.

Sousa, R. B., Sobreira, H., & Moreira, A. P. (2023). A systematic literature review on long-term localization and mapping for mobile robots. *Journal of Field Robotics*, *40*(5), 1021–1054.

Sun, J., Zhao, J., Hu, X., Gao, H., & Yu, J. (2023). Autonomous navigation system of indoor mobile robots using 2D lidar. *Mathematics*, *11*(6), 1455.

Tóth, Z., Caruana, R., Gruber, T., & Loebbecke, C. (2022). The dawn of the AI robots: Towards a new framework of AI robot accountability. *Journal of Business Ethics*, *180*(3), 727–748.

Tran, Q.-K., & Ryoo, Y.-J. (2025). Multi-sensor fusion framework for reliable localization and trajectory tracking of mobile robot by integrating UWB, odometry, and AHRS. *Biomimetics*, *10*(3), 345.

Tsintotas, K. A., Bampis, L., & Gasteratos, A. (2022). The revisiting problem in simultaneous localization and mapping: A survey on visual loop closure detection. *IEEE Transactions on Intelligent Transportation Systems*, *23*(8), 11016–11036.

Wang, H., He, L., Zhang, S., Bai, R., & Wang, Y. (2025). Mobile Robot path planning considering obstacle gap features. *Applied Sciences*, *15*(11), 5678.

Wei, Z., Wang, S., Chen, K., & Wang, F. (2025). ROS-based navigation and obstacle avoidance: A study of architectures, methods, and trends. *Sensors*, *25*(13), 5090.

Wu, Y., Zhang, Y., Zhu, D., Deng, Z., Sun, W., Chen, X., & Zhang, J. (2023). An object SLAM framework for association, mapping, and high-level tasks. *IEEE Transactions on Robotics*, *39*(4), 2794–2813.

Yan, Y., Zhang, B., Zhou, J., Zhang, Y., & Liu, X. (2022). Real-time localization and mapping utilizing multi-sensor fusion and visual–IMU–wheel odometry for agricultural robots in unstructured, dynamic and GPS-denied greenhouse environments. *Agronomy*, *12*(8), 1740.

Yang, L., Tao, Y., Li, M., Zhou, J., Jiao, K., & Li, Z. (2026). LiDAR-inertial SLAM method integrated with visual QR codes for indoor mobile robots. *Scientific Reports*, *16*, 145.

Yang, Z., Li, J., Yang, L., Li, P., & Xia, G. (2023). Path planning and collision avoidance methods for distributed multi-robot systems in complex dynamic environments. *Mathematical Biosciences and Engineering*, *20*(5), 8815–8842.

Yao, C., Ge, Y., Shi, G., Wang, Z., Yang, N., Zhu, Z., Wei, H., Zhao, Y., Wu, J., & Jia, Z. (2024). TAIL: A terrain-aware multi-modal SLAM dataset for robot locomotion in deformable granular environments. *IEEE Robotics and Automation Letters*, *9*(2), 1670–1677.

Zhang, B., Peng, Z., Zeng, B., & Lu, J. (2024). 2DLIW-SLAM: 2D LiDAR-inertial-wheel odometry with real-time loop closure. *Measurement Science and Technology*, *35*(7), 075402.

Zhang, X. Y., Abd Rahman, A. H., & Qamar, F. (2023). Semantic visual simultaneous localization and mapping (SLAM) using deep learning for dynamic scenes. *PeerJ Computer Science*, *9*, e1513.

Zhao, J., Liu, S., & Li, J. (2022). Research and implementation of autonomous navigation for mobile robots based on SLAM algorithm under ROS. *Sensors*, *22*(11), 4143.

Zheng, S., Wang, J., Rizos, C., Ding, W., & El-Mowafy, A. (2023). Simultaneous localization and mapping (SLAM) for autonomous driving: Concept and analysis. *Remote Sensing*, *15*(5), 1278.

Zhou, Z., Feng, X., & Di, S. (2023). A LiDAR mapping system for robot navigation in dynamic environments. *IEEE Transactions on Intelligent Vehicles*, *8*(5), 2851–2864.

Zhu, J., Li, H., & Tao, Z. (2023). Camera, LiDAR, and IMU based multi-sensor fusion SLAM: A survey. *Tsinghua Science and Technology*, *28*(5), 880–899.

Zohaib, M., Ahsan, M., Khan, M., & Iqbal, J. (2023). A featureless approach for object detection and tracking in dynamic environments. *PLOS ONE*, *18*(1), e0280001.

11 Adaptive and Learning Robots

11.1 INTRODUCTION

Rule-based robots operate using instructions from rules and programming that they were designed with. Adaptive robots can change or adapt their programming when they encounter new conditions or situations. The traditional method of robots working well within closed environments but fail when they come across new or changing situations has led researchers to develop systems that learn over time how to complete specific tasks (Zhu et al., 2025).

Learning robots allow robots to use information gathered from their sensors and experiences to improve how they respond to commands over time (Shahid et al., 2022). Instead of robots requiring explicit step-by-step instructions for a task, they use experiential information to adapt to uncertainty.

A robot can learn to perform specific tasks through trial and error. When an action is taken by a robot, sensors allow it to learn from that action using reward-based metrics. If the robot performed well during the action, it will be provided with a reward. If it performed poorly, it would not receive a reward or would receive a penalty. The robot will learn to take actions that will achieve the most reward over time, which will result in it learning to complete a desired task (Radosavovic et al., 2023).

Learning builds off most of the topics covered in previous chapters. Robots need to perceive their environment using the Perception chapter information, control systems to actuate (Control Systems), and navigate to enable movement (Navigation). Learning allows robots to use information acquired from their environment and change how it reacts to certain actions.

Chapter 5 covers adaptive and learning robots. Adaptive robots can alter their responses based on what data they receive. Learning robots use iterations to improve how they complete tasks. Topics include how robots learn from experiences (Scheikl et al., 2023), reinforcement learning, ways to train AI, and implementing learning in real-world environments.

11.2 FOUNDATIONS OF LEARNING IN ROBOTICS

Learning systems allow robots to become better at a task with experience instead of only using hard-coded behaviours (Tang et al., 2024). Robotics has traditionally been rule-based, but using machine learning allows robots to change behaviours by analysing data and learning from their mistakes. This allows robots to complete more complicated tasks that are closer to real-world applications (Radosavovic et al., 2023).

The most common learning method used in perception tasks is supervised learning, which uses labelled data like images to learn how to recognise objects/features.

DOI: 10.1201/9781003758501-15

This type of learning does not work as well for decision-making because we often do not know the correct answer to tell the robot.

Reinforcement learning allows robots to learn through their own experiences. The robot is not told the correct answer, but it is allowed to try an action and receive feedback. The robot will receive a reward or penalty based on its performance and will learn to optimise for the highest reward over time (Panda et al., 2024; Tang et al., 2024).

Experience is defined by the sensorimotor interaction an agent receives from its environment (Saviolo et al., 2022). Each time a robot interacts with the world, it receives experience. A robot can learn from experience by repeating actions and learning from its mistakes (Yang et al., 2025).

Since the world is uncertain, robots must generalise from their experiences to know what to do in new situations (Scheikl et al., 2023).

11.3 MODELLING LEARNING IN ROBOTIC SYSTEMS

When studying how robots learn from experience, it is helpful to have a visual model that describes this concept (Milani et al., 2023). The RL loop model is a helpful visual representation to show how robots take in information from their environment and update their responses as they learn more over time (Orr & Dutta, 2023).

The diagram starts with an agent which symbolises the robot or learner. This agent exists within an environment which consists of what the robot can perceive or sense. These two things work in tandem repeatedly (Shahid et al., 2022).

In each cycle, the agent begins by observing the current state. A state is the data that the robot can understand about its situation. For instance, where it is located, what its sensors are reading, and what is around it. After getting this information, the agent will choose an action to take (Milani et al., 2023).

Once the action has been chosen, the robot carries out that action and the environment returns a reward and a new state (Ibrahim et al., 2024). A reward is simply a score used to show how good of a decision the action was. If the robot did something it should, it will receive a positive reward. If it did something it shouldn't have, it will receive a negative reward.

Finally, the agent updates using this information. Specifically, it changes its policy. Policy is how the robot decides what action to take when it encounters a state (Orr & Dutta, 2023). As the agent learns over time and maximises reward, it should ideally start performing better due to updating its policy.

Figure 11.1 represents a schematic diagram demonstrating this closed loop. Knowing the state of the environment, the agent performs an action and is given a reward based on its behaviour. It then updates its policy and the process repeats. From this RL loop, learning occurs through ongoing interaction, with feedback being used to allow the agent to learn from its experiences.

11.4 CASE STUDY: GOOGLE EVERYDAY ROBOT PROJECT

Google Everyday Robot project is one of the robots that learns from real-world experience (Tang et al., 2024). It learns how to accomplish various everyday tasks through machine learning instead of being explicitly programmed to do so. Everyday tasks can

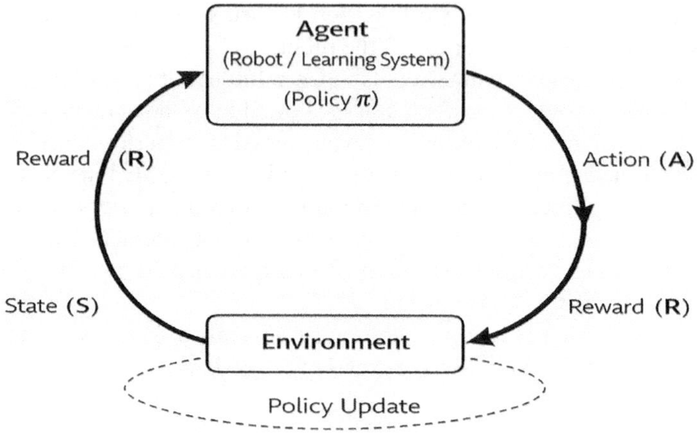

FIGURE 11.1 Reinforcement learning loop in robotics.

include picking up objects, opening doors, wiping tables, etc. Tasks like these are easy for humans but hard for robots because of their inherent uncertainty (Han et al., 2023).

The robot learns from its experience through trial and error by using reinforcement learning. It will choose an action given its state in the environment. The robot will receive a reward or penalty and update its knowledge based on its action (Shahid et al., 2022). Over time, the robot will learn how to act in its environment without being told.

One challenge with this approach is that it takes much time and data to learn from real-world experience. The robot may try to learn a task for years without ever succeeding at it (Han et al., 2023). Learning is usually done with simulators and then transferred to the real world.

Another issue is knowing when to explore new actions and when to exploit what is already known (Han et al., 2023; Tang et al., 2024). Learning is also harder when there is noise in the world around us (Radosavovic et al., 2023).

11.5 REINFORCEMENT LEARNING FUNDAMENTALS

In reinforcement learning (RL), robots learn by interacting with their environment and receiving feedback in the form of rewards or penalties (Tang et al., 2024). The robot isn't told how to behave; instead, it discovers its behaviour based on which actions lead to which rewards or penalties (Han et al., 2023). This approach is ideal for robotics problems because it's often challenging or impossible to know every detail about the robot's surroundings (Hohenfeld et al., 2022).

The four main principles of reinforcement learning are states, actions, rewards, and policies (Han et al., 2023):

State is the robot's current situation in its environment. For example, a state might include the location of nearby objects, the robot's speed, or the status of its sensors (Miranda et al., 2022). All decisions are made based on the robot's state.

Action refers to any movement or changes the robot can make. Movements can range from simply rotating in place to picking up an object (Shahid et al., 2022). Basically, an action is anything that changes the state.

Reward is the feedback that the robot receives after taking an action. Rewards can be positive or negative and are used to indicate whether the robot is behaving as it should (Cai et al., 2023). The robot's goal is to learn to earn as many rewards as possible.

Policy is the strategy that the robot employs to determine what action it should take based on its current state (Miranda et al., 2022). As the robot learns, its policy will change to choose the actions that are most likely to earn rewards in the future (Shahid et al., 2022; Tang et al., 2024).

Reinforcement learning is about trying different actions and refining your policy based on the rewards you receive. Given enough time, the robot will learn behaviours that are suited to its environment and the task it is trying to accomplish (Hohenfeld et al., 2022).

These four RL principles are fundamental to learning how to program robots. Make sure your students understand states, actions, rewards, and policies before moving on to more complex topics.

11.6 TRAINING AND SIMULATION ENVIRONMENTS

Robots can be trained using reinforcement learning, but to do this, they must experience large amounts of data. Training experiences can be expensive or time-consuming to collect in the real world (Ju et al., 2022). Simulation environments are therefore an integral part of learning-based robot design pipelines (Cao et al., 2022). Before robots interact with the real world, they can try again in simulation. This lets engineers know that robots will behave safely and as expected in many situations (Scheikl et al., 2023).

In simulation, the robot receives observations from its simulated environment. The environment can be anything – from an empty room to a crowded street with other cars and objects – that can be modelled in software. Developers can also model real-world physics, such as engine limits, collisions, and other constraints (Wiberg et al., 2023). With reinforcement learning, the robot can try taking thousands, if not millions, of different actions during training and learn from each one without harming equipment or bystanders (Wu et al., 2025).

Training entirely in simulation has many benefits. Simulation can be sped up to allow a robot to experience years of driving in a matter of days. Simulation also prevents wear and tear on the robot's hardware and prevents it from getting smashed into a wall while it is still learning (Cao et al., 2022).

Unfortunately, there is often a gap between how well a model performs in simulation versus how it performs. Engineers call this the reality gap. The reasons for this gap can range from sensor noise to different physical constants or even unforeseen conditions in the real world (Ju et al., 2022). To account for this reality gap, engineers will use better-designed simulations and even test their model with real-world data.

Environment realities also depend on whether data is available for collecting. In reinforcement learning, we allow the agent to collect its own data instead of providing labelled examples. Interaction is used to train the robot and collect data. Reward signals can come from anything, but how the robot perceives its environment will affect how it acts within that environment. Ensuring that the robot experiences enough of these states is critical to how it will behave at test time (Wu et al., 2025).

Simulation and real-world testing are combined by engineers to build robust training pipelines for adaptive robots (Cao et al., 2022). Robotic systems can learn from mistakes quickly and safely using these pipelines, enabling the practical use of learning-based robotics.

11.7 EXPLORATION AND EXPLOITATION

An exploration-exploitation trade-off is one of the problems that need to be addressed when developing reinforcement learning algorithms (Moradi et al., 2024). Exploration and exploitation are approaches to learning that determine how a robot should select its actions. Exploration consists of choosing actions that appear new or unknown to the system, while exploitation refers to choosing actions that are known to yield great rewards (Nguyen & Han, 2023). A robot learning system needs to choose when to exploit what it has learned and when to explore new opportunities.

The problem with learning through exploration is that it can be random and without direction. If a robot employed only exploration to learn about its environment, it could take a long time. On the other hand, if it solely exploits what it knows, it may never find the best policy.

Balancing exploration with exploitation is beneficial to the robot. When exploiting, a robot applies what it knows about its environment to choose an action that it predicts will have the best reward (Moradi et al., 2024). For instance, if a robot is learning how to walk through a room; through exploration, it determines how to move forward. Once it knows how to walk forward, it starts to exploit and will continually move forward until it learns another way to reach its goal.

Since the robot is aware of how to move forward, it will get the most reward by repeatedly taking that action. There is a balance to strike between exploration and exploitation. Too much exploration can cause a system to take longer to learn, and its behaviour may seem erratic. Too much exploitation can cause your system to get stuck in a local maximum and never learn the best policy (Tran & Kim, 2022).

Robotics tends to complicate the exploration-exploitation trade-off because physical systems can be damaged through exploration. Getting a robot to move around randomly can cause it to collide with walls or hurt itself. We must restrict how much a robot will explore when learning about its environment. Most reinforcement learning systems use some form of exploration and exploitation to solve this problem. A robot may explore a room randomly at first to gain information about its environment and slowly start exploiting what it knows about the best path through the room. Martinez-Baselga et al. (2023) showed how exploration and exploitation can be combined to teach robots to walk.

11.8 LEARNING IN UNPREDICTABLE ENVIRONMENTS

Robots interacting with real environments may face uncertain, stochastic, or non-stationary behaviours (Nguyen & Han, 2023). Real environments can differ from simulated or controlled ones in numerous ways and can change while the robot is operating within them (Radosavovic et al., 2023). Learning how to safely and reliably navigate these conditions can be difficult (Saviolo et al., 2022; Zhu et al., 2025).

Environmental variability: Learning systems need to deal with changes in lighting conditions, the presence of objects or even changes to surface properties that could affect how the robot senses and moves through the environment. A learning agent may need to generalise what it has learned in the past to be deployed in a wide range of situations it has never experienced.

Noise and uncertainty: Robot sensors can experience interference, inaccuracy, or lag that leaves the robot uncertain about what is around it or what is happening in its environment. Learning needs to take place while accounting for this noise and uncertainty so that the robot will not take risky actions based on incorrect interpretations of its sensors.

Non-stationarity: Another challenge with real-world environments is that they are rarely static (Zhu et al., 2025). Objects can move unexpectedly, new obstacles can appear, and the overall conditions that the robot encountered when it first began its task can change.

Exploration vs. exploitation: The trade-off between trying new things and sticking with what is known to work must be balanced against safety and reliable operation in physical systems (Panda et al., 2024). Exploration must be limited or performed in such a way that the system can't get hurt or break from trying new things.

Generalisation: Many real-world deployment environments will differ from the conditions the robot was trained under (Radosavovic et al., 2023). Learning algorithms that rely on memorisation of past experiences will not generalise well to new environments. Instead, agents should aim to learn a representation of the world that lets it apply past experiences to new settings.

By taking these issues into account when designing learning systems, robots can be made to better adapt to a wider variety of conditions that they may encounter in real environments. This is useful to allow learned behaviours to be deployed in places outside of simulated environments or environments that the robot was specifically trained to operate in (Nguyen et al., 2023; Zhu et al., 2025).

11.9 PRACTICAL INSIGHT: Q-LEARNING IN PYTHON

Q-learning is a basic reinforcement learning technique that allows a robot (agent) to learn what action to take at each moment by trial and error (Orr & Dutta, 2023). The agent develops a Q-table that it uses to predict the expected reward when taking an action while in each state. As the agent learns from its environment, it updates the Q-table.

Essentially, the agent will constantly do this:

1. Perceive its environment
2. Take an action
3. Get a reward
4. Learn from the experience

Here's a simple implementation of Q-learning:

Listing 11.1: Q-learning agent in Python

```
import numpy as np
import random

# Define environment (simple grid world)
states = 5
actions = 2  # 0 = left, 1 = right

# Initialise Q-table
Q = np.zeros((states, actions))

# Parameters
alpha = 0.1    # learning rate
gamma = 0.9    # discount factor
epsilon = 0.2 # exploration rate

# Training loop
for episode in range(1000):
    state = random.randint(0, states-1)

    for step in range(10):
        # Choose action (exploration vs exploitation)
        if random.uniform(0,1) < epsilon:
            action = random.randint(0, actions-1)
        else:
            action = np.argmax(Q[state])

        # Simulate environment
        next_state = max(0, min(states-1, state + (1 if action ==
1 else -1)))
        reward = 1 if next_state == states-1 else 0

        # Q-learning update
        Q[state, action] = Q[state, action] + alpha * (
            reward + gamma * np.max(Q[next_state]) - Q[state,
action]
        )

        state = next_state

print("Learned Q-table:")
print(Q)
```

This example illustrates how an agent learns to navigate to a goal state by updating its Q-values using reward feedback. The agent initially wanders aimlessly but begins to exploit after several iterations. Q-learning shows how robots can learn from their environments without being explicitly taught. The agent will learn an optimal policy by updating its behaviour as it experiences new things

11.10 CHALLENGES IN LEARNING ROBOTICS SYSTEMS

Possible challenges of learning are related to data efficiency (Tang et al., 2024). Exploration during learning might lead to unsafe actions, simulation-to-reality gap, limited generalisation (Han et al., 2023); computational complexity of RL algorithms; and lack of interpretability and transparency (Vouros, 2022).

Learning from scratch through trial-and-error interactions with real-world environments can require lots of samples, which makes it sample inefficient. In particular, when learning by taking actions in a real robot, this sample inefficiency translates to slow learning which might be physically slow (each iteration takes minutes or longer to complete), costly (if the robot gets damaged during training), and incapable of repetition (unlike simulations, we cannot just reset the environment and try again after a crash).

Systems that learn from interactions with their environment may perform unsafe actions during exploration. Balancing safe operation and exploration is an ongoing research topic going into areas like safe reinforcement learning. We don't want the system to damage itself, its environment, or nearby humans while it is learning.

Learning in a simulated environment is often more sample efficient, allowing one to train efficiently before deploying to the real world. However, sim-to-real gaps can result in poor performance once a system has learned the desired behaviour in simulation.

A robot that learned to navigate in one building may fail if placed in a new building due to limited ability to generalise to new environments.

Most reinforcement learning algorithms tend to require significant compute resources and memory to run, which may be restrictive for some real-world robots.

Learning systems are often considered "black boxes," meaning it is hard to know what caused the system to make the decision that it did. Debugging these types of systems can become more difficult. Humans trust what they understand. If the system can't explain why it made the decisions that it did, it will be hard for humans to trust it.

11.11 ETHICAL AND REAL-WORLD CONSIDERATIONS

Learning approaches present unique ethical and social considerations since robot behaviour emerges from experience rather than being preprogrammed (Raptis et al., 2025). Issues include safety and accountability during and after learning.

Safety during learning: Since many reinforcement learning approaches involve exploration, robots may try unsafe actions that harm themselves or others. Safe exploration methods are needed to constrain actions to safe subsets (Raptis et al., 2025; Singh & Singh, 2024).

Predictability: Continual adaptation can make a robot's behaviour harder to predict than programmed behaviour. This may make learned robots less reliable for safety-critical applications like medical or autonomous vehicles.

Bias: A robot's experience may not be comprehensive or representative of its entire operating domain. If it learns from biased experience, it may make unfair decisions or unsafe choices when faced with situations outside its training data (Varsha, 2023).

Responsibility: If a robot learns its behaviour from data rather than being explicitly programmed, who is responsible for its actions? Developers? Operators who monitored learning? The larger organisation?

Trust: Users need to know that learning robots will perform safely, fairly, and reliably before allowing them into everyday environments (Singh & Singh, 2024).

11.12 CONCLUSION

This chapter delved into the realm of robotics systems that learn from experience, moving away from rigidly programmed robots bound by predetermined rules. It discovered that reinforcement learning provides a robust framework for facilitating such learning, with concepts like states, actions, rewards, and policies laying the groundwork for training machines. The reinforcement learning loop model offered us a structured approach to understanding the learning process. By examining the Everyday Robot project by Google, we witnessed how learning unfolds through real-world tasks. Additionally, we explored training methods and environments, including simulation-based approaches, and tackled the exploration vs. exploitation dilemma. As confronted challenges like data requirements, safety concerns, and generalisation hurdles, it became evident that deploying learning systems in the real world demands careful consideration. Ethical implications also emerged as an essential aspect to ponder when designing autonomous learning systems. In conclusion, learning empowers robots to think and act more intelligently and flexibly than ever before, but not without introducing new challenges along the way.

11.13 QUESTIONS FOR DISCUSSION

1. Why is reinforcement learning a natural fit for robotics?
2. What are the primary elements of RL?
3. How do robots know when to explore versus exploit?
4. What issues can occur when applying learning from simulation to the real world?
5. What ethical issues are raised by learning based robots?

11.14 EXERCISES

Exercise 1: Implement Q-Learning

Implement a simple Q-learning agent as described above and show how it learns to improve its performance over time. This should be straightforward given our description.

Exercise 2: Recognise RL Components

Describe the states, actions, and rewards associated with a robotic task of your choosing.

Exercise 3: Exploration vs. Exploitation

Show how performance varies when using different exploration methods.

Exercise 4: Reflect on Ethics

Why is it dangerous to let robots learn from interaction with the real world? What responsibilities do learning robot scientists have to protect the public when experimenting with learning robots?

REFERENCES

Cai, M., Xiao, S., Li, J., & Kan, Z. (2023). Safe reinforcement learning under temporal logic with reward design and quantum action selection. *Scientific Reports, 13*, 1842.

Cao, H., Theile, M., Wyrwal, F. G., & Caccamo, M. (2022). Cloud-edge training architecture for sim-to-real deep reinforcement learning. *2022 IEEE/RSJ International Conference on Intelligent Robots and Systems (IROS)*, 2716–2723.

Han, D.-K., Mulyana, B., Stanković, V., & Cheng, S. (2023). A survey on deep reinforcement learning algorithms for robotic manipulation. *Sensors, 23*(7), 3452.

Hohenfeld, H., Heimann, D., Wiebe, F., & Kirchner, F. (2022). Quantum deep reinforcement learning for robot navigation tasks. *IEEE Access, 10*, 20462–20481.

Ibrahim, S., Mostafa, M., Jnadi, A., Salloum, H., & Osinenko, P. (2024). Comprehensive overview of reward engineering and shaping in advancing reinforcement learning applications. *IEEE Access, 12*, 95540–95572.

Ju, H., Juan, R., Gomez, R., Nakamura, K., & Li, G. (2022). Transferring policy of deep reinforcement learning from simulation to reality for robotics. *Nature Machine Intelligence, 4*(12), 1120–1135.

Martinez-Baselga, D., Riazuelo, L., & Montano, L. (2023). Improving robot navigation in crowded environments using intrinsic rewards. In *2023 IEEE International Conference on Robotics and Automation (ICRA)* (pp. 11699–11705). IEEE.

Milani, S., Topin, N., Veloso, M., & Fang, F. (2023). Explainable reinforcement learning: A survey and comparative review. *ACM Computing Surveys, 55*(10), 1–38.

Miranda, V. R. F., Neto, A. A., Freitas, G., & Mozelli, L. A. (2022). Generalization in deep reinforcement learning for robotic navigation by reward shaping. *IEEE Transactions on Industrial Electronics, 69*(12), 13616–13626.

Moradi, M., Zhai, Z.-M., Panahi, S., & Lai, Y.-C. (2024). Adaptive network approach to exploration–exploitation trade-off in reinforcement learning. *Chaos, 34*(12), 123103.

Nguyen, H., Andersen, R. S., Boukas, E., & Alexis, K. (2023). Uncertainty-aware visually-attentive navigation using deep neural networks. *The International Journal of Robotics Research, 42*(14), 1254–1284.

Nguyen, H. D., & Han, K. (2023). Safe reinforcement learning-based driving policy design for autonomous vehicles on highways. *International Journal of Control, Automation and Systems, 21*(12), 3930–3943.

Orr, J., & Dutta, A. (2023). Multi-agent deep reinforcement learning for multi-robot applications: A survey. *Sensors, 23*(7), 3456.

Panda, P., Sahoo, D., & Sahoo, D. (2024). Deep reinforcement learning for real-time robotic control in dynamic environments. *2024 International Conference on Computational Intelligence and Networks (CINE)*, 1–6.

Radosavovic, I., Xiao, T., Zhang, B., Darrell, T., Malik, J., & Sreenath, K. (2023). Real-world humanoid locomotion with reinforcement learning. *Science Robotics, 8*(75), eade2255.

Raptis, E. K., Kapoutsis, A. C., & Kosmatopoulos, E. B. (2025). Agentic LLM-based robotic systems for real-world applications: A review on their agenticness and ethics. *Frontiers in Robotics and AI, 12*, 1543210.

Saviolo, A., Frey, J., Rathod, A., Diehl, M., & Loianno, G. (2022). Active learning of discrete-time dynamics for uncertainty-aware model predictive control. *IEEE Transactions on Robotics, 38*(6), 3876–3895.

Scheikl, P. M., Tagliabue, E., Gyenes, B., Wagner, M., Dall'Alba, D., Fiorini, P., & Mathis-Ullrich, F. (2023). Sim-to-real transfer for visual reinforcement learning of deformable object manipulation for robot-assisted surgery. *IEEE Robotics and Automation Letters, 8*(2), 968–975.

Shahid, A. A., Piga, D., Braghin, F., & Roveda, L. (2022). Continuous control actions learning and adaptation for robotic manipulation through reinforcement learning. *Autonomous Robots, 46*(5), 739–758.

Singh, K. D., & Singh, P. (2024). Fog cloud computing and IoT integration for AI enabled autonomous systems in robotics. *EAI Endorsed Transactions on AI and Robotics, 4*(1), e5.

Tang, C., Abbatematteo, B., Hu, J., Chandra, R., Martín-Martín, R., & Stone, P. (2024). Deep reinforcement learning for robotics: A survey of real-world successes. *Annual Review of Control, Robotics, and Autonomous Systems, 7*, 451–482.

Tran, V. M., & Kim, G.-W. (2022). Cooperative deep reinforcement learning policies for autonomous navigation in complex environments. *IEEE Access, 10*, 132414–132431.

Varsha, P. S. (2023). How can we manage biases in artificial intelligence systems: A systematic literature review. *International Journal of Information Management Data Insights, 3*(1), 100168.

Vouros, G. (2022). Explainable deep reinforcement learning: State of the art and challenges. *ACM Computing Surveys, 55*(9), 1–39.

Wiberg, V., Wallin, E., Fälldin, A., Semberg, T., Rossander, M., Wadbro, E., & Servin, M. (2023). Sim-to-real transfer of active suspension control using deep reinforcement learning. *Robotics and Autonomous Systems, 162*, 104366.

Wu, M., Wu, L., Wu, Y., Huang, W., Fan, H., Hu, Z., Geng, H., Li, J., Ying, J., Yang, L., Chen, Y., & Dong, H. (2025). SimLauncher: Launching sample-efficient real-world robotic reinforcement learning via simulation pre-training. *2025 IEEE/RSJ International Conference on Intelligent Robots and Systems (IROS)*.

Yang, Y., Yu, H., Lou, X., Liu, Y., & Choi, C. (2025). Attribute-based robotic grasping with data-efficient adaptation. *IEEE Transactions on Robotics, 41*(1), 250–268.

Zhu, Y., Wan Hasan, W. Z., Harun Ramli, H. R., Norsahperi, N. M. H., Kassim, M. S. M., & Yao, Y. (2025). Deep reinforcement learning of mobile robot navigation in dynamic environment: A review. *Sensors, 25*(11), 3456.

Part V

Humans, Ethics, and Systems Thinking

12 Human–Robot Interaction

12.1 INTRODUCTION

As robots expand their presence outside of factories and warehouses and into environments designed for and dominated by humans, their interactions with people will become increasingly important (Panagou et al., 2023; Su et al., 2023). Human–robot interaction (HRI) is the discipline focused on developing robot systems that can interact with humans naturally and effectively; in other words, communicate, cooperate, and coexist with humans (Zhang et al., 2024). This area prioritises user experience over raw technical capabilities.

Humans may also have expectations for robots that go beyond these basic needs. We typically want robots to behave according to accepted norms; gesture and speak when addressed, maintain a comfortable physical distance from people, and act predictably (Su et al., 2023). These attributes may become even more important when robots work in close cooperation with people, such as caregivers in hospitals and patient homes, teachers and support staff in educational settings, and customer service representatives (Stock-Homburg & Kegel, 2025).

You have already learned about several of the building blocks of HRI. Preceptors can help robots sense when humans are present and allow them to receive signals, we can communicate such as speech or hand gestures (Su et al., 2023). Controllers and navigators can allow robots to move through and physically interact with spaces shared with humans. Learners can allow robots to modify their behaviour when interacting with humans based on how people react to them (Park & Whang, 2022). Using these concepts, robots can meaningfully interact with humans (Zhang et al., 2024).

Since humans are involved, interaction is not always easy to design. People might react to robots in unexpected ways. Culture, context, and other individual factors may impact how we expect robots to behave (Abbo et al., 2025). The same physical robot could be viewed as helpful or annoying; autonomous or creepy (Stock-Homburg & Kegel, 2025).

This chapter will begin our introduction to the principles of HRI. You will learn how robots can detect, recognise, and socially/emotionally interact with people in their environments (Park & Whang, 2022). This includes how robots can speak to people and physically interact with them, as well as topics like trust and ethics related to robots that interact with humans (Panagou et al., 2023).

12.2 FOUNDATIONS OF HUMAN–ROBOT INTERACTION

HRI is the field of study dedicated to understanding how humans and robots interact with one another (Safavi et al., 2024). Robotics heavily involves psychology, graphic design, and human–computer interaction in order to meet users' needs and expectations and mimic human behaviour (Zhang et al., 2024).

DOI: 10.1201/9781003758501-17

One of the central guidelines of HRI is user-centred design (UCD). Robots should behave predictably and intuitively and be easy to use. Robot systems should not make users change the way they live or work around a robot but rather be accessible and simple to use.

There is also a need for robots to balance between being interactive and automated. Robots are automated in that they do not need input from humans to complete tasks, but they must also be able to interact with humans by accepting input and providing output (Panagou et al., 2023).

Humans communicate with robots through input such as speech, gestures, and facial expressions. Robots then use sensors and programmed instructions to make sense of that information and infer the user's intent (Su et al., 2023).

Additionally, the context of interaction may also need to be taken into consideration, such as whether the robot is interacting with a human in a hospital setting versus a retail setting (Panagou et al., 2023; Park & Whang, 2022).

Through these considerations, HRI aims to facilitate easier and more comfortable interactions with robots and increase social acceptance of robotics (Rossi et al., 2024).

12.3 MODELLING INTERACTION IN ROBOTIC SYSTEMS

There are many ways to model interaction between robots and humans. One way is to view interaction as a continuous loop (Safavi et al., 2024; Su et al., 2023). HRI loop represents flow of information between human and robot and how each effect other's behaviour.

Interaction starts with human input. Human input can be speech, gestures, facial expressions, movement or any other way of conveying intent (Su et al., 2023). Human expresses what they want from the robot (e.g. request, command, question or simple social gesture). Robot's perception system recognises human input (Zhao et al., 2025). Robot has multiple sensors that allow it to sense environment, like microphone, camera, and motion sensors. Robot will use perception algorithms to interpret input. For example, speech recognition will convert spoken words to commands. Vision system can be used to recognise gestures or faces (Su et al., 2023).

Robot will then output response (Habibian et al., 2024). This could be action taken by the robot or robot expressing something through speech, movement or changing the lights (Safavi et al., 2024). Response should meet goals of the context and match what human would expect from robot. Feedback from human comes as they see the response and do something about it. Human could provide another input and continue the cycle (Habibian et al., 2024).

Figure 12.1 models this interaction as a cycle: input from the human(s) results in perception by the robot, which leads to a response and feedback by the human(s). There is now potential for continuous conversation/exchange with adaptation on both sides. Figure 12.1 emphasises that interaction is cyclical and continuous. Successful back and forth relies on accurate perception, appropriate response and clear feedback.

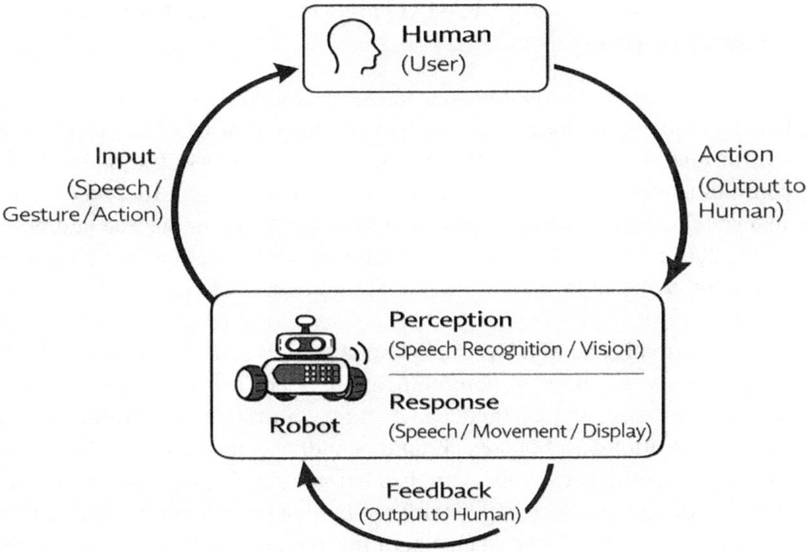

FIGURE 12.1 Human–Robot interaction loop.

Modelling this cycle allows designers to consider how to build systems that not only converse but have a pleasant user experience.

12.4 CASE STUDY: PEPPER ROBOT IN UK CARE HOMES

Pepper is a social humanoid robot used to assist in UK care homes. Pepper aims to aid socialisation and improve resident welfare. Pepper's capabilities include face detection, basic emotion detection, conversation, and responding via speech and gestures (Mihailov & Wangmo, 2025). Pepper reminds residents of daily activities, reminds them to drink water, and assists with socialisation to prevent loneliness.

There are three concepts of HRI featured within this case study. Firstly, trust and likeability are important considerations in HRI. If Pepper is perceived as nice and can hold users' attention, residents' mood may increase, and loneliness may decrease. Secondly, usability. Interfaces need to be usable and easily accessible. As older adults in residential homes may not be as comfortable or used to using new technologies, interactions should be straightforward and mimic natural modes of communication, speech-based is one (Lee et al., 2022).

Users can grow attached to robots on an emotional level. This does not mean that robots can feel emotion, but Pepper detects user input and displays the output most suited to the situation it detects/has learned over time (Mihailov & Wangmo, 2025).

Limitations include that robots cannot be used as a replacement for human contact. Residents may choose to only socialise with Pepper over care workers. Pepper can only do as its programmed, if it doesn't sense its environment accurately, it will provide an output based on what it thinks the users' input was.

12.5 COMMUNICATION MODALITIES IN HUMAN–ROBOT INTERACTION

Humans and robots can communicate using several modalities (Safavi et al., 2024). Modality refers to a communication mechanism through which two or more entities transfer information from one to another and make it understood (Su et al., 2023). Each communication modality enables the robot to understand human input provided in a specific format and then allows it to generate an output that addresses the provided input (Su et al., 2023). Supporting multiple modalities enables human–robot communication to be more natural, flexible, and accessible (Urakami & Seaborn, 2022).

Voice interaction is one of the primary modalities to consider (Lima et al., 2022). With speech as the medium of communication, users can naturally interact with robots using their voice. Speech input can contain commands, questions, or allow for free-form conversations. Robot speech understanding will interpret what is said by using speech recognition. The robot can then output speech using speech synthesis to relay information or simply converse with the human user. Voice is a great medium of communication with robots when operating hands-free is necessary. This can occur when your hands are occupied, as is common in some healthcare and service industries (Lima et al., 2022).

Gesture is another common modality and often complements voice interactions (Peral et al., 2022). Rather than saying something, people will point, use their hands to explain direction or distance, and wave. If robots can see the user through a vision system, gestures can be recognised and acted upon by robots. Gestures can provide a stronger cue than speech in some scenarios and can be used on their own when speech cannot be used (Peral et al., 2022).

Displays, lights, and facial expressions are another form of communication modality. Robots can output visual cues to users to relay information about what they are doing, what they want users to do, or what is going on with the robot. Simple indicators can help users understand a robot's status or what will happen next (Safavi et al., 2024).

Interaction can also happen through multiple modalities at once (Su et al., 2023). This is known as multimodal interaction and allows users to interact with the system using voice, gestures, or visual cues. Responses can also use multiple modalities to improve overall robustness and usability of the system.

Selection of appropriate modalities will depend on the use case. Consideration should be given to context, user needs, and the environment among many other things (Urakami & Seaborn, 2022). Supporting a variety of modalities allows robots to better interact with users in a natural, intuitive, and familiar manner. This allows for richer and more accessible HRI (Safavi et al., 2024; Su et al., 2023).

12.6 PROXEMICS AND PHYSICAL INTERACTION

Proxemics defines humans' behaviour when they occupy space when interacting, which includes body distances, positioning/orientation, and movement (Samarakoon et al., 2022). Proxemics has a significant impact on HRI to make a robot feel more natural (Neggers et al., 2024). Robots must maintain an appropriate amount of distance between themselves and others and move around without causing confusion or fear.

Distance between two agents is an important category in proxemics (Leichtmann et al., 2022). Humans navigate according to their desired personal space or intention, whether it be personal, social, or public distance. Robots need to observe this to ensure the user does not feel uneasy with a robot coming too close or too far away (Xu et al., 2023). Carefully designing distance strategies will ensure comfortable proxemics.

Approach distance and positioning/orientation also impact Human–Robot proxemics (Xu et al., 2023). If a robot faces a person or how it approaches can play a factor in how the robot is perceived. Approaching someone from the front can feel more natural than approaching from behind to maintain the line of sight. Addressing these concerns can create a more comfortable user experience (Neggers et al., 2024).

Motion strategy and smoothness also affect how people feel around robots. Certain movements can make people feel uneasy and unaware of what a robot may do next. By implementing predictable motions and movement around people that are instinctual to humans can improve HRI. Robots need to move smoothly and properly to fit the context of human surroundings.

The environment also plays a role in proxemics (Leichtmann et al., 2022). Since proxemics itself is contextual, it is important to design robots with public interaction in mind. Robots may be able to get closer to humans in a crowded public bus versus a hospital healthcare robot. Culture also plays a large role in proxemics and should also be considered when designing robots (Samarakoon et al., 2022).

Designing robots with proxemics in mind can allow for more natural interactions with humans. It will allow users to feel more comfortable around robots and communicate more effectively, allowing for smoother HRIs in real-world applications.

12.7 DESIGNING FOR APPROACHABILITY AND TRUST

Robots must appear approachable, predictable, and trustworthy to humans if they are to succeed in human environments (Adeniyi, 2025). People are more inclined to approach and interact with robots if they look safe, understandable, and responsive (Pinney et al., 2022). HRI designers should therefore prioritise approachability and trustworthiness.

Appearance seems to be the single most critical factor. Humans judge robots based on their physical appearance, including shape, size, facial features, and more (Kim, 2024). Friendly-looking robots with simple, understandable features are often viewed as more approachable (Pinney et al., 2022). Robotic features that appear too complex or human-like can make users uneasy, especially if the robot does not act as expected.

Behavioural cues also strongly influence user perceptions of predictability. Robots should behave in ways that humans can understand and anticipate their actions. If a robot is predictable, people will know what to expect and will feel more confident around the machine (Adeniyi, 2025). Movement patterns, reaction time, and responsiveness can all affect how predictable a robot appears (Mårell-Olsson et al., 2025).

Users should also never feel lost or unsure about what a robot is doing. Transparency can help humans understand a robot's actions and make the machine appear more intelligent (Zerilli et al., 2022). Speakers, visual signals, and movement can provide user feedback to reinforce intent. If a robot says what it's doing or where it's going next, people will have an easier time interacting with it.

Responsiveness is another important factor in perceived trustworthiness. Robots should always respond quickly and correctly to user actions. Response time is critical, and slow or inaccurate reactions can quickly fluster users and destroy trust in the system (Zhang & Doyle, 2023).

Trust must also be maintained over time. Users need time to develop positive feelings towards a robot, and sudden changes in behaviours can undo prior progress (Adeniyi, 2025). A trustworthy robot will behave consistently across contexts and applications.

Ensuring approachability and trust requires consideration of the robot's appearance, predictability of behaviour, transparency of action, responsiveness to user input, and consistency over time (Pinney et al., 2022). By meeting user expectations in each of these areas, robots can provide positively human-like experiences in real-world environments (Łukasik & Gut, 2025).

12.8 CULTURAL AND EMOTIONAL CONSIDERATIONS

Culture plays a major role in HRI design. Humans of different backgrounds may view robot behaviour differently (Kim, 2024). Emotional expectations can also vary between people and impact interactions with robots. Account for all these factors is essential when designing robots to be used in various contexts (Urakami & Seaborn, 2022).

Culture affects factors such as social norms, expected verbal and non-verbal communication, and general technology expectations. This might include how direct or indirect it is appropriate to be, how much eye contact and physical distance to maintain, or how formal you should be when communicating (Urakami & Seaborn, 2022). If you were to design a robot to operate in one culture without considering these variables, it may engage in socially unacceptable behaviour when placed in a different context. Culture can deeply impact not just what your robot says but also how it moves and behaves (Kim, 2024).

Emotional considerations are another thing to keep in mind. When interacting with robots, people tend to assign them intelligence and emotion that they may not actually have. Users can have very different reactions to robot behaviour based on whether or not they perceive the robot to be empathetic. Users might come to trust a robot and view it as more competent if it smiles and speaks in a soothing tone, for example (Park & Whang, 2022).

There is a difference between a robot mimicking empathy and having actual emotions. Robots can be designed to recognise human emotions and react in what it thinks is an appropriate manner. They are not sentient beings and do not feel emotions like we do (Park & Whang, 2022). Your robot can acknowledge a user's frustration and respond in a helpful manner without feeling sympathy for them (Su et al., 2023). Understanding that your robot has no emotions can prevent confusion on behalf of the user.

Good emotional design should not trick your users into thinking your robot feels emotions but should provide comfortable and appropriate interactions. A robot used for healthcare may use a gentle, soothing voice to convey care for their patients without having to worry about displaying false emotions.

You should also consider the context your robot will be used in. Emotional design and cultural expectations can change based on application. Different contexts can include eldercare, education, logistics, customer service, and more (Lechuga Redondo et al., 2023). By creating robots with multiple context-specific interaction models, you can ensure that they properly adapt to each situation they may encounter.

Humans of different cultures may perceive your robots differently, and customers have many expectations when it comes to robot emotions. By keeping your design decisions within these users' comfort zones, you can create robotic technology that is more accessible to all kinds of people.

12.9 PRACTICAL INSIGHT: DESIGNING A ROBOT INTERACTION

Real-world interactions often benefit from specifying the robot dialogue behaviour beforehand. One way to do so is by designing a dialogue flow. This means explicitly specifying what the user says and how the robot should respond at each step of the conversation. Anticipating possible conversation branches can also allow the designer to minimise misunderstanding and create responses that are transparent, predictable, and familiar to the user (Tisserand et al., 2024).

Dialogue flows start by identifying the user's intention. When humans interact with a reception robot, for example, they can have several intentions: they can ask for directions, they can ask for help or information. The robot needs to detect the user's intent and react accordingly. The designer can map the set of user messages to each potential intent and to actions performed by the robot in response (Belcamino et al., 2026).

Interaction then flows through various states such as greeting the user, exchanging information, requesting clarification, or ending the conversation. It should be noted that states represent points in the conversation, while transitions are dependent on both user input and system reaction. With properly defined states and transitions, complex conversations can be tackled while limiting confusion (Tisserand et al., 2024).

Listing 12.1: Example dialogue flow for a robot receptionist

```
def robot_interaction(user_input):
    if "hello" in user_input:
        return "Hello! How can I assist you today?"

    elif "directions" in user_input:
        return "Please tell me where you would like to go."

    elif "office" in user_input:
        return "The office is on the second floor. Take the lift
on your left."
```

```
    elif "thank you" in user_input:
        return "You're welcome! Have a great day."

    else:
        return "I'm sorry, I didn't understand that. Could you
please repeat?"
```

One application of using conversation flows as a part of robot User interface (UI) is shown in this example. User gives some input to the robot and based on what the user said it replies. Of course, in a more realistic situation, a more sophisticated approach would be taken and might understand small variations in input as well as keeping track of conversation longer. The interaction also provides feedback so users know what the robot understands and how the conversation will continue. Clarity of communication, manners, and predictable behaviour also play an important role in user experience. Designing conversation flows allows you to design these interactions so that they work well and feel natural to the user. It's a very practical way of mapping out how a robot will communicate in its intended environment.

12.10 CHALLENGES IN HUMAN–ROBOT INTERACTION

HRI suffers from many complications and setbacks as opposed to traditional technical systems. Because of this field's connection to human beings, real-life environments that systems must operate in, and the complexity of human behaviour itself (Safavi et al., 2024; Su et al., 2023), HRI is much more difficult to design and subsequently test (Urakami & Seaborn, 2022). Miscommunication can easily be made by both parties: robots can take users' input the wrong way and act according to wrong metrics. For example, this applies to speech recognition or gesture recognition software. Users can also misunderstand what a robot is trying to do if its behaviour does not match its intentions. Both parties must successfully take the other's input and convey their messages to maintain interaction. This is why not all communication methods used by humans work for HRI. Users might develop beliefs that robots possess intelligence levels or emotions that allow them to competently comprehend contextual information, intention, subtext, and subtle social cues (Park & Whang, 2022). If robots are unable to act on these implications, it can lead to user errors or loss of trust in the technology. As a result, users need to have manageable expectations for the robot.

Robots may display a form of artificial empathy that can sometimes seem as if they can socially interact with their users. However, this is not true, as these behaviours are scripted by pre-defined rules and responses (Park & Whang, 2022). If these responses do not match how a user acts or expects the robot to act, it can create a perceivably awkward interaction.

Robots should also be able to detect what context a user is in and act appropriately. Whether a certain action is acceptable for a robot is going to vary between users and environments (Mårell-Olsson et al., 2025). Because of this, an HRI system will have to be able to detect its surroundings and the user it is interacting with to properly respond to them (Safavi et al., 2024). The limitation of current technology is another large factor. HRI systems are limited by their processing power, sensitivity of detection systems, and understanding of language.

12.11 ETHICAL AND REAL-WORLD CONSIDERATIONS

Ethics become an issue with HRI because robots are being used in more socially aware or emotionally charged situations (Abbo et al., 2025). Robots' actions and communication can cause shifts in users' emotions, choices, and relationships. This means they need to be built carefully to avoid doing harm wherever possible (Macalupu et al., 2025).

Manipulation of emotions: Robots shouldn't push emotional reactions. Some artificially intelligent robots are programmed to emulate empathy or friendship. Users might start to feel more attached or become emotionally dependent on these robots. There is concern about whether humans should be replaced with programmed machines that don't fully understand these emotions themselves (Mihailov & Wangmo, 2025).

Becoming too dependent: Robots are great when you need them, but they can be harmful if used too often. If someone spends all their time with a social robot in their care home, they can become too dependent. This reduces human contact, which could be detrimental to their long-term health.

Privacy: Just like with normal tech, robots will acquire data about the user. This can come from listening to speech, observing behaviours, and even reading emotional reactions. It's important that proper security measures are kept avoiding breaching any data (Weerarathna et al., 2023).

Knowing when to trust: Users should know the limitations of a robot and its strengths.

Replaceability: If robots are going to take the place of humans in caretaking or social functions, what does that mean for society? There are questions about how we care for others, provide jobs, and interact on a personal level (Mihailov & Wangmo, 2025).

12.12 CONCLUSION

This chapter discussed robots interacting with humans, taking into consideration social and emotional aspects. The robot system design needs to be suitable for the users, and it needs to be usable so that it will be accepted. In designing robotic systems and robots, the following HRI fundamentals need to be taken into consideration: Human-Centred Design of Robots, Modality of Communication and Proxemics (Robot) related Physical Interaction. Interaction loop diagrams showed how humans and robots exchange information bidirectionally. Pepper, a social robot implementation in UK care homes provided case study experiences on trust, usability, and reaction for robots used in sensitive environments. Cultural, emotional and expected levels, and robot/system behaviour affected interactions. Misunderstandings in communication, unrealistic expectations, and robot/system emotional mismatches were challenges identified with the case study. Privacy, dependency, and trust were discussed from an ethical perspective. Robots need to be able to interact with humans socially and emotionally. HRI must be taken into consideration when designing robots.

12.13 DISCUSSION QUESTIONS

1. What factors determine trust between humans and robots?
2. How do different modalities impact interaction quality?
3. What is proxemics and why does it matter in HRI?
4. What are some considerations when designing robots for emotion-sensitive environment?
5. Can robots be empathetic? Should they try to be empathetic?

12.14 EXERCISES

Exercise 1: Designing Dialogues.

Design a dialogue flow based on a robot receptionist answering multiple requests.

Exercise 2: Contextual Comparison.

Compare and contrast two contexts of human–robot interaction (e.g., eldercare, education, retail).

Exercise 3: Evaluating Interactions.

Choose an interaction you have seen or done with a robot. What were some strengths? Weaknesses?

Exercise 4: Ethics Reflection.

Should robots be allowed to "build" relationships with users on an emotional level? Why or why not?

REFERENCES

Abbo, G. A., Belpaeme, T., & Spitale, M. (2025). Concerns and values in human-robot interactions: A focus on social robotics. *International Journal of Social Robotics.*

Adeniyi, D. (2025). The role of trust in human-robot interaction: A perspective on factors influencing trust development. *International Research Journal of Innovations in Engineering and Technology*, *9*(1), 112–120.

Belcamino, V., Kilina, M., Carfì, A., Seidita, V., Mastrogiovanni, F., & Chella, A. (2026). Factored reasoning with inner speech and persistent memory for evidence-grounded human–robot interaction. *Frontiers in Robotics and AI.*

Habibian, S., Alvarez Valdivia, A., Blumenschein, L. H., & Losey, D. P. (2024). A survey of communicating robot learning during human–robot interaction. *The International Journal of Robotics Research.*

Kim, R. Y. (2024). Anthropomorphism and human-robot interaction. *Communications of the ACM*, *67*(2), 20–23.

Lechuga Redondo, M. E., Niewiadomski, R., Rea, F., Incao, S., Sandini, G., & Sciutti, A. (2023). Comfortability analysis under a human–robot interaction perspective. *International Journal of Social Robotics*, *15*(5), 1003–1021.

Lee, O., Lee, H., Park, A., & Choi, N. G. (2022). My precious friend: Human-robot interactions in home care for socially isolated older adults. *Clinical Gerontologist*, *45*(5), 1186–1200.

Leichtmann, B., Lottermoser, A., Berger, J., & Nitsch, V. (2022). Personal space in human-robot interaction at work: Effect of room size and working memory load. *ACM Transactions on Human-Robot Interaction*, *11*(3), 1–28.

Lima, M. R., Wairagkar, M., Gupta, M., Rodriguez y Baena, F., Barnaghi, P., Sharp, D., & Vaidyanathan, R. (2022). Conversational affective social robots for ageing and dementia support. *IEEE Transactions on Cognitive and Developmental Systems*, *14*(4), 1307–1321.

Łukasik, A., & Gut, A. (2025). From robots to chatbots: Unveiling the dynamics of human–AI interaction. *Frontiers in Psychology*, *16*, 1432103.

Macalupu, V., Miller, E., Martin, L., & Caldwell, G. (2025). Human–robot interactions and experiences of staff and service robots in aged care. *Scientific Reports*, *15*, 1312.

Mårell-Olsson, E., Bensch, S., Hellström, T., Alm, H., Hyllbrant, A., Leonardson, M., & Westberg, S. (2025). Navigating the human–robot interface—Exploring human interactions and perceptions with social and telepresence robots. *Applied Sciences*, *15*(3), 1452.

Mihailov, E., & Wangmo, T. (2025). Does humanness matter? An ethical evaluation of sharing care work with social robots. *Science and Engineering Ethics*, *31*(3), 41.

Neggers, M. M. E., Belgers, S., Cuijpers, R. H., Ruijten, P. A. M., & IJsselsteijn, W. (2024). Comfortable crossing strategies for robots. *International Journal of Social Robotics*, *16*(1), 77–96.

Panagou, S., Neumann, W. P., & Fruggiero, F. (2023). A scoping review of human–robot interaction research towards Industry 5.0 human-centric workplaces. *International Journal of Production Research*, *61*(7), 2201–2226.

Park, S., & Whang, M. (2022). Empathy in human–robot interaction: Designing for social robots. *International Journal of Environmental Research and Public Health*, *19*(3), 1764.

Peral, M., Sanfeliu, A., & Garrell, A. (2022). Efficient hand gesture recognition for human–robot interaction. *IEEE Robotics and Automation Letters*, *7*(4), 9475–9482.

Pinney, J., Carroll, F., & Newbury, P. (2022). Human-robot interaction: The impact of robotic aesthetics on anticipated human trust. *PeerJ Computer Science*, *8*, e942.

Rossi, A., Holthaus, P., Moros, S., Lakatos, G., Andriella, A., Scheunemann, M. M., & Maris, A. V. (2024). Trust, acceptance and social cues in human-robot interaction (SCRITA). *International Journal of Social Robotics*, *16*(3), 789–798.

Safavi, F., Olikkal, P., Pei, D., Kamal, S., Meyerson, H., Penumalee, V., & Vinjamuri, R. (2024). Emerging frontiers in human–robot interaction. *Journal of Intelligent & Robotic Systems*, *109*(2), 35.

Samarakoon, S., Muthugala, M. V. J., & Jayasekara, A. (2022). A review on human–robot proxemics. *Electronics*, *11*(16), 2559.

Stock-Homburg, R. M., & Kegel, M. (2025). Ethical considerations in customer–robot service interactions: Scoping review, network analysis, and future research agenda. *International Journal of Social Robotics*.

Su, H., Qi, W., Chen, J., Yang, C., Sandoval, J., & Laribi, M. (2023). Recent advancements in multimodal human–robot interaction. *Frontiers in Neurorobotics*, *17*, 1123456.

Tisserand, L., Stephenson, B., Baldauf-Quilliatre, H., Lefort, M., & Armetta, F. (2024). Unraveling the thread: Understanding and addressing sequential failures in human-robot interaction. *Frontiers in Robotics and AI*, *11*, 1420382.

Urakami, J., & Seaborn, K. (2022). Nonverbal cues in human–robot interaction: A communication studies perspective. *ACM Transactions on Human-Robot Interaction*, *11*(3), 1–32.

Weerarathna, I. N., Raymond, D., & Luharia, A. (2023). Human-robot collaboration for healthcare: A narrative review. *Cureus*, *15*(11), e49321.

Xu, X., Liying, L., Khamis, M., Zhao, G., & Bretin, R. (2023). Understanding dynamic human-robot proxemics in the case of four-legged canine-inspired robots. In *2023 IEEE International Conference on Robotics and Automation (ICRA)* (pp. 1234–1240). IEEE.

Zerilli, J., Bhatt, U., & Weller, A. (2022). How transparency modulates trust in artificial intelligence. *Patterns*, *3*(2), 100513.

Zhang, Y., & Doyle, T. (2023). Integrating intention-based systems in human–robot interaction: A scoping review of sensors, algorithms, and trust. *Frontiers in Robotics and AI, 10*, 1245678.

Zhang, Y., Kassem, K., Gong, Z., Mo, F., Ma, Y., Kirjavainen, E., & Häkkilä, J. (2024). Human-centered AI technologies in human–robot interaction for social settings. In *Proceedings of the International Conference on Mobile and Ubiquitous Multimedia* (pp. 1–6). ACM.

Zhao, I., Leung, A. Y. M., Huang, Y., & Liu, Y. (2025). A social robot in home care: Acceptability and utility among community-dwelling older adults. *Innovation in Aging*, *9*(1), igae108.

13 Ethical and Responsible Robotics

13.1 INTRODUCTION

Ethics dictates how we should act and what is right or responsible behaviour. As robots gain autonomy, become more intelligent, and increasingly proliferate, ethical considerations about their behaviour become more important (Hwang, 2024). Robots are operating in homes, hospitals, public spaces, and battlefields. They see, hear, perceive, make decisions, and take actions that impact human safety, privacy, and well-being (Elendu et al., 2023). Robots may also act independently of a human guiding their actions, creating challenges around who is responsible when something goes wrong (Tóth et al., 2022).

Autonomous robots take actions in the physical world using data, programmed instructions, and learned behaviours, their decisions have consequences (Elendu et al., 2023). For this reason, ethics also considers how robots are designed, trained, and deployed. Responsibility for robot actions is difficult to assign, especially when systems malfunction or behave unexpectedly.

Robotic systems can encounter ethical dilemmas throughout their entire lifecycles. For instance, if a robot utilises machine learning, it could exhibit bias from its training data and cause harm to individuals (Mali & Bhosale, 2025). Many parts of robotics feed into ethical behaviour, including perception, learning, and human–robot interaction (HRI), earlier chapters provide foundational concepts for understanding these topics.

This chapter presents ethical frameworks commonly used in robotics and discusses ethics topics including bias, transparency, accountability, and regulation (Hwang, 2024).

13.2 FOUNDATIONS OF ETHICS IN ROBOTICS

Robotics ethics deals with issues regarding morally correct conduct in the design, manufacture, and utilisation of robots (Huang et al., 2023; Licardo et al., 2024). Ethics is becoming increasingly important in robotics as robots are able to make more decisions independently from humans and act upon them. Ensuring that robots are used safely and responsibly requires ethical principles (Callari et al., 2024).

Ethics, in general, involves questions about what is considered right or wrong and how one should act. Since robots can make decisions without human intervention, they can affect how people live and work. There are many ethical concerns with robotics that need to be considered (Elendu et al., 2023; Hwang, 2024).

The first issue is autonomous decision-making. Robots can sense their environment and make decisions based on what they perceive. Sometimes, it is not clear how a robot will make its decision or who is actually in control of the robot (Hwang, 2024).

DOI: 10.1201/9781003758501-18

The second issue is societal impact. Robots can have many positive impacts on society, but they can also cause problems. They can replace human jobs or enable mass surveillance, which can negatively impact privacy.

The third concern with robotics ethics is the responsibility of engineers and designers. When engineers create a system, they typically focus on how it will operate when everything goes right. However, they should also consider how it may be used maliciously or how it could fail (Arora et al., 2025; Callari et al., 2024; Elendu et al., 2023; Radanliev, 2025).

Some ethical considerations can conflict with one another. It might be more efficient to allow a robot to violate someone's privacy, but that would be unethical. Ethics seeks to provide principles for people to follow when they encounter these dilemmas (Huang et al., 2023).

13.3 ETHICAL FRAMEWORKS FOR ROBOTICS

Ethical judgments often consider established ethical frameworks (Hwang, 2024). An ethical framework describes how humans view right or wrong behaviour. We can use ethical frameworks to help assess potential robotics solutions (Winfield et al., 2025). Popular ethical frameworks include consequentialism, duty/deontology, and virtue ethics.

Consequentialism bases ethical judgements on the impact of an action. When using consequentialism, one makes ethical decisions by choosing the action with the best consequence. In robotics, it may involve deciding which action maximises the number of lives saved. For example, when building an autonomous system, designers may use utilitarian logic to reduce overall damage.

Duty and deontology determine ethicality by rules. When using duty or deontology, some actions are ethical or unethical independent of other factors (Hwang, 2024). Robotic decisions may include hardcoded rules that the system cannot break. Some companies will not allow users to opt out of privacy protocols, no matter the potential benefits (Fink et al., 2023).

Virtue ethics determines ethicality by identifying virtuous characteristics that drive behaviour. Instead of assessing individual actions, virtue ethics considers human nature. Virtue ethics in robotics ensures a system acts like a "moral patient" or virtuous person (Arora et al., 2025). We may design systems with virtues like fairness and account for responsibility when developing new technology (Ye, 2025). For example, we can design autonomous systems to mimic human ethical decision-making behaviour (Arora et al., 2025).

Different ethical frameworks can provide opposing solutions to the same problem (Winfield et al., 2025). For example, surveillance cameras may be ethical from a utilitarian perspective because they will save lives, but not from a deontological perspective because they invade privacy (Hwang, 2024). Understanding how each framework differs allows us to see the difficulties in problem-solving ethics (Tóth et al., 2022).

Using these models, designers can examine robot systems through various lenses. This allows for ensuring decisions aren't purely technical but ethically derived as well.

13.4 CASE STUDY: AI-POWERED DRONES IN SURVEILLANCE AND MILITARY USE

AI-powered drones being used for surveillance and military strikes represent some of the most ethically complicated use cases today (Caballero-Martin et al., 2024). Drones can move throughout space autonomously or be directed by a user, perceive their surroundings in real-time and, in some cases, make decisions based on their programming with little to no human input (Caballero-Martin et al., 2024). These benefits do not come without potential for issues with responsibility, privacy, and collateral damage occurring (Geetha et al., 2025).

Facilities being monitored by drones are one example of these systems being utilised. Drones with AI systems can survey large areas, track movements, and gather additional data to feed into their algorithms (Sholademi, 2024). This allows for more manpower to be allocated to areas that need it and keeps civilians safer. Of course, there is still risk to individuals' privacy. If civilians are being always tracked by drones, there is a higher potential for that data to be abused.

Another prime example of AI drones being abused is with militaries using them for their own gain (King, 2024). AI systems can be used to help soldiers with drone strikes or reconnaissance. Life or death decisions can be made by autonomous technology with little oversight (Nasehi, 2025). When an issue like this occurs, who is responsible? The people who created and maintained the AI system? The person operating the machinery it is installed on. The entire organisation that made the decision to utilise these systems?

Collateral damage is another risk with AI drones being used in battle or outdoor spaces (King, 2024). AI systems are only ever going to see the world as they are programmed. Pictures and sensors can feed information to these systems, but they only perceive the world as ones and zeros (Nasehi, 2025). Even if the AI can recognise something as a threat, there is no way for technology to know if it made the correct choice. Friendly fire incidents could occur, or violence could be drawn simply because of misinformation (Ahmad et al., 2025). Operating in outdoor environments can also lead to mistakes. Just because a drone can navigate an area, it does not mean that it will be able to account for things like weather conditions.

Utilitarianism and deontology provide interesting lenses through which to look at this scenario. A utilitarian would think of whether drones lead to less casualties, while deontology would have individuals question if assassinating someone with a drone is even right in the first place (Riesen, 2022). If we were to look at this through the lens of virtue ethics, we would hold the creators of these AI systems to a higher standard of moral and ethical responsibility.

13.5 BIAS IN ALGORITHMS AND DATA

Bias and discrimination in algorithms and training data are some of the largest ethical issues in robotics today (Pagano et al., 2023). Robots that make their own decisions with machine learning (ML) have inherent risks of bias. Bias in systems refers to when they output systematically unfair or incorrect results (Jui & Rivas, 2024). Bias can affect perception, decision-making, and HRI (Larsson et al., 2023).

Bias can stem from both training data and the algorithm itself. Training a system using datasets which do not account for every scenario can cause issues when it is exposed to something that it does not recognise. For example, if a vision system is constantly run indoors while being trained, it will not know how to interpret images when taken outside (Pagano et al., 2023).

Bias can also be caused by the algorithm that processes inputs itself. If certain pathways are preferred or features are weighted more heavily than others, a system may miscategorise things or treat certain things differently than others (Jui & Rivas, 2024).

Bias can cause issues with each individual subset of robotics. Biased perception can cause wrong image or pattern recognition. Biased decision-making can result in unfair or incorrect actions taken by a robot. Biased HRI can cause robots to treat certain users differently than others.

Avoiding bias can be done by using diverse datasets to train systems, ensuring systems are tested in the environments they will operate in, and consistently monitoring system performance (Pagano et al., 2023). Designing systems with transparency can also allow for areas of bias to be spotted (Jui & Rivas, 2024).

By limiting bias, robots and robotic systems can operate more fairly, be trusted more by their users, and interact with more diverse groups of people (Larsson et al., 2023).

13.6 EXPLAINABLE AI AND TRANSPARENCY

Explainability in AI systems (XAI) is the process by which we can make the behaviour of an intelligent system understandable by humans (Vouros, 2022). Transparency is a similar, but separate property, focused around making the inner workings of a system visible (Methnani et al., 2024). Explainability and transparency have become major topics as robots have begun utilising unintuitive machine learning models.

Most state-of-the-art AI systems today are considered "black boxes." This is especially true of deep learning-based systems, which, while capable of amazing performance, offer little to no insight into why a particular decision was made (Frasca et al., 2024). In robotics, this can lead to users not trusting a robot to complete a task, or dangerous scenarios if the robot can't explain its actions (Methnani et al., 2024).

While explainability can provide benefits to trust from a user perspective, it can also help with debugging if a robot makes a mistake (Ye et al., 2023). Knowing why a decision was made can allow a developer to track down where the underlying model made an error.

The key benefit of explainable AI in robots, however, is accountability. If a robot is making decisions that have consequences in the real world, we must be able to examine how those decisions were made to ascribe responsibility and guarantee certain standards are upheld (Frasca et al., 2024).

Often, there is a trade-off between the complexity of a model and its interpretability (Frasca et al., 2024). We can design for explainability by using more interpretable models and providing explanations to users. Not only will this allow your robot to be more trustworthy, but safer and more responsible as well.

13.7 LEGAL AND SOCIAL IMPLICATIONS

Robotic systems may have legal and social implications that must be considered as they become more autonomous and integrated into society (Tóth et al., 2022). Legal frameworks and social norms can impact robotic systems in various ways.

Accountability is a significant legal concern with robotic systems. If something goes wrong, such as a robotic system making a harmful decision, it may be difficult to assign blame (Tóth et al., 2022). Who is responsible for the system's actions – the developer, manufacturer, operator, or the organisation deploying it? Determining liability can be difficult, especially when the system uses machine learning that evolves over time (Pham, 2025).

Regulation is another important consideration for robotic systems. Governments and regulatory bodies must establish rules and guidelines to ensure that robotic systems are safe, reliable and deployed ethically (Licardo et al., 2024). This may include regulations around testing and certification, as well as standards for operation. However, regulation can struggle to keep up with technological advancements, leading to gaps in oversight.

Privacy concerns are also relevant to robotic systems. Many collect and process vast amounts of data, including personal data and behaviour. Ensuring that data is managed securely and in compliance with privacy laws such as data protection legislation is crucial to protecting people's rights (Pham, 2025).

Socially, robotics can also impact employment and society (Licardo et al., 2024). Automation has the potential to replace jobs but create new ones. Depending on how automation is implemented, there may be significant changes to the workforce. Questions about fairness and access may also arise if robotic systems become essential to society (Díaz Rodríguez et al., 2023).

Public trust is another social factor. People must trust robotic systems for widespread adoption to occur. This means that they must feel safe around them, believe they can perform their intended tasks reliably and that ethical considerations have been thoroughly considered. Building public trust requires effective design, communication, and responsible deployment practices (Ye et al., 2023).

These examples show that robotics is not just a technical discipline but also a social one (Licardo et al., 2024). Legal and social implications must be considered when designing, deploying and regulating robotic systems. Collaboration between engineers, policymakers, and society is required to ensure that robotic systems are used safely and for the benefit of everyone (Díaz Rodríguez et al., 2023).

13.8 ETHICAL RISK ASSESSMENT IN ROBOTICS

Ethical risk assessment refers to processes that systematically identify, analyse, and address potential ethical concerns that may arise when developing and deploying robotic systems (Winfield et al., 2025). When robots are deployed in real-world settings, they can have profound impacts on their users and the environments they operate in. As robots gain greater autonomy and capabilities, it is increasingly important to consider these risks before they occur to ensure that robots operate safely, fairly, and responsibly (Huang et al., 2023).

The ethical risk assessment process starts with risk identification, which involves determining how your robotic system may affect users, society, and the environment (Winfield et al., 2025). Risks can include safety concerns, privacy risks, unintended biases in decision-making, or other negative consequences of robot behaviour (Elendu et al., 2023). Consideration should be made for both intended and unintended uses of your system (Dodig-Crnkovic et al., 2023).

Once risks have been identified, the next step is to evaluate each risk's potential impact and likelihood of occurring. Impact refers to the severity of consequences should the risk occur. Some failures may be almost negligible if they occur, while others could cause serious injury or damage. Likelihood is how probable it is that the risk will occur. Example failures should be considered to have low likelihood but very high impact. This evaluation process allows you to prioritise which risks need more attention than others (Winfield et al., 2025).

The final stage of ethical risk assessment is to mitigate each risk or provide a design response. This involves developing strategies to eliminate or reduce the likelihood of each risk occurring (Huang et al., 2023). Solutions can range from design changes to improve the system to mitigating factors that can reduce the impact of the failure should it occur. Examples of responses include adding constraints to robot actions, improving data quality, or adding safe human oversight. Consideration should be made for ethical risks during the design process instead of addressing them after the system has already been built (Dodig-Crnkovic et al., 2023).

Ethical risk assessment should not only be performed before deployment, but throughout the system's entire lifecycle. Monitoring your system will allow you to catch new ethical risks that may come up as you update your robot's software or hardware, or if it gets deployed in new environments (Winfield et al., 2025). Continued evaluation also allows you to ensure that your system always operates safely (Elendu et al., 2023; Huang et al., 2023).

Ethical risk assessment should also consider different perspectives. This can include the technical/medical perspectives, but also legal and social perspectives. By exploring an issue from many angles, you can make sure that your team is making well-informed decisions that consider all important stakeholders (Callari et al., 2024).

Ethical risk assessment allows roboticists to identify and address ethical concerns proactively. By considering these issues during the design process, you can help ensure that your system doesn't negatively impact users and society and that it provides benefit to the greatest number of people (Winfield et al., 2025).

13.9 PRACTICAL INSIGHT: ETHICAL EVALUATION FRAMEWORK

Ethics guidelines need actionable tools that enable people to assess specific real-world systems. An ethics evaluation framework provides a repeatable process you can follow when analysing a robot before releasing it. By checking your system across several criteria, you can help mitigate risks and responsibilities while highlighting the impacts your technology may have on society (Díaz Rodríguez et al., 2023).

Such frameworks generally ask you a series of questions along several axes, such as security, fairness, transparency, accountability, and more (Elendu et al., 2023;

Radanliev, 2025). This enables you to clearly see where your system stands ethically (Tóth et al., 2022).

Below is an example of one simple ethical checklist you can use to help evaluate your systems:

Listing 13.1: Ethical assessment checklist

```
def ethical_assessment(system):
    checklist = {
        "safety": "Does the system minimise risk of harm?",
        "privacy": "Is user data protected and handled
responsibly?",
        "fairness": "Does the system avoid bias and unequal
outcomes?",
        "transparency": "Can decisions be explained and
understood?",
        "accountability": "Who is responsible for system
actions?",
        "reliability": "Does the system perform consistently?"
    }

    for key, question in checklist.items():
        print(f"{key.upper()}: {question}")
```

The questions above can serve as a template or a checklist when evaluating robotics systems. The important part is that each question prompts further introspection about potential harms.

When asking such questions in practice, it can be useful to have multiple people from a diverse set of backgrounds doing the analysis. An engineer or designer may come to different conclusions than a policy expert or a future user of the system.

Ethics isn't a checkbox; it can continue to evaluate an existing system as it continues to change over time. For example, you may find that the system impacts more users or is being used in different ways than originally intended.

Frameworks like these can help translate high-level ethics concepts to real systems. This helps developers apply ethics to the design of their technology to build responsible and transparent robotics.

13.10 CHALLENGES IN ETHICAL ROBOTICS

Even with established ethical frameworks and principles, implementing ethical decision-making processes into robots is difficult (Dodig-Crnkovic et al., 2023; Winfield et al., 2025). Some ethical challenges include conflicts in values or trade-offs, meaning different stakeholders may value privacy, safety, efficiency, or equality more, creating difficulty in deciding what to prioritise when designing robots (Raptis et al., 2025). Uncertainty in the real world is another obstacle designers might face. Robots often operate with incomplete or noisy data, making it difficult for them to know how their actions will impact ethics or ensure that the robot will always act ethically (Winfield et al., 2025). Ambiguous situations with no morally correct action can also arise, requiring the robot to balance different ethical principles.

Culture and background can affect how people view ethical behaviour. Because robots will be used in many places, they must account for differences in culture and legal ethics (Tóth et al., 2022; Ye, 2025). Additionally, there are limitations on the technology's side. A robot may not know how to identify human values or safety concerns and can be led into biased or discriminatory actions based on its programming or data it learned from (Arora et al., 2025; Radanliev, 2025; Raptis et al., 2025). Ethics also need to be implemented and enforced after the robots are released by regulations, policies, or monitoring, making it another challenge. Ethics in robotics is complicated and depends on the situation, and it is important that many perspectives are considered when involving ethics.

13.11 ETHICAL AND PHILOSOPHICAL DEBATE

Robotics ethics and philosophy examine how society reacts to automation and intelligent machines (Huang et al., 2023). Here are some ethical issues people have with robots:

Robots taking human jobs: Robots automate processes and make them faster and more efficient. While automation can be beneficial, too much of it can cause humans to be phased out of decision-making processes.

Lethal autonomous weapons: Should machines be allowed to decide who lives and who dies? Some believe taking humans out of the equation will decrease mistakes and fatalities, but others disagree (Sehrawat, 2025).

Privacy: Some robots can survey the world around them indefinitely. This could be used to closely monitor people's private spaces and actions, which could be subject to abuse if not properly regulated (Corrêa et al., 2022).

Keeping humans "in the loop": Humans should always have a degree of control over robots. As robots become more intelligent, there is some concern about keeping humans involved in their decision-making processes (Sehrawat, 2025; Tóth et al., 2022).

Moral agency of robots: Some philosophers believe that robots are not moral agents and will never be (Hwang, 2024; Tóth et al., 2022). While robots can be programmed with a moral compass, they will never possess true consciousness and therefore cannot understand it.

Robot ethics and philosophy will become increasingly more important as we move forward. Experts from all fields must work together to help combat these issues (Arora et al., 2025).

13.12 CONCLUSION

The chapter focused on ethical and social issues that modern robotic systems bring along with the fast pace of new technological developments. It emphasised the importance of responsible robotics design and decision-making alongside technical advancements. Throughout the chapter, three main schools of ethics are presented – consequentialism, deontology, and virtue ethics. Each provides unique approaches for assessing robotic systems regarding the potential good or harm they can impose on society.

Algorithmic bias, transparency, accountability, and new legal considerations for actions made by machines autonomously without human intervention were presented. A case study about AI-powered drones was discussed to emphasise these topics' real-world impact better. Autonomous drones can lead to severe consequences if used incorrectly, posing safety, privacy, and human life risks.

Robot ethics evaluation frameworks were provided to aid designers and other stakeholders with the ethical risk assessment of robotics systems before deployment. Other significant difficulties when implementing ethics into robotics design include issues with conflicting values, uncertainty, value misalignment, and cross-cultural differences.

Broader ethical concerns and topics of discussion in the robotic ethics community were presented, such as ethical use of autonomous weapons and limits to machine autonomy. The chapter drives home the point that robotics is as much about technical achievements as it is about morals. Robot systems need to be safe, fair, transparent, and upholding of human values to be successful.

13.13 DISCUSSION QUESTIONS

1. Which ethical framework(s) do you think is best suited for analysing robots? Why?
2. Give an example of how bias in algorithms could have real-world consequences.
3. Why is explainability important in autonomous systems?
4. Who should be responsible if a robotic system causes harm?
5. Should lethal autonomous weapons ever be allowed?

13.14 EXERCISES

Exercise 1: Ethical Risk Assessment

Choose or imagine a robotic system you would like to design. Perform an ethical risk assessment. Describe at least two risky behaviours that your future robot could exhibit and discuss how you would prevent these.

Exercise 2: Framework Application

Choose or imagine an applicable real-world robotics situation. Apply consequentialism, deontology, and virtue ethics to this scenario.

Exercise 3: Bias Analysis

Choose or imagine an existing robotic system. Identify where bias can be introduced in the data/system used to make decisions by the robot.

Exercise 4: Debate Exercise

Should lethal autonomous weapons ever be legal? Discuss. Provide multiple ethical perspectives on the topic.

REFERENCES

Ahmad, I., Ahmad, L., Irshad, N., & Talha, M. (2025). Artificial intelligence in autonomous weapon systems: Legal accountability and ethical challenges. *Journal of Engineering, Science and Technological Trends*, *7*(1), 45–58.

Arora, A., Marshall, A., Arora, A., & McIntyre, J. R. (2025). Virtuous integrative social robotics for ethical governance. *Discover Artificial Intelligence*, *3*, 7.

Caballero-Martin, D., López-Guede, J. M., Estévez, J., & Graña, M. (2024). Artificial intelligence applied to drone control: A state of the art. *Drones*, *8*(7), 327.

Callari, T., Vecellio Segate, R., Hubbard, E.-M., Daly, A., & Lohse, N. (2024). An ethical framework for human-robot collaboration for the future people-centric manufacturing. *Technology in Society*, *75*, 102579.

Corrêa, N., Galvão, C., Santos, J., Pino, C., Pinto, E. P., Barbosa, C., Massmann, D., Mambrini, R., Galvao, L., & Terem, E. (2022). Worldwide AI ethics: A review of 200 guidelines and recommendations for AI governance. *Patterns*, *3*(7), 100588.

Díaz Rodríguez, N., Ser, J., Coeckelbergh, M., López de Prado, M., Herrera-Viedma, E., & Herrera, F. (2023). Connecting the dots in trustworthy artificial intelligence: From AI principles, ethics, and key requirements to responsible AI systems and regulation. *Information Fusion*, *99*, 101866.

Dodig-Crnkovic, G., Holstein, T., Pelliccione, P., & Thavarasa, J. (2023). Future intelligent autonomous robots, ethical by design – Lessons learned from autonomous cars ethics. *Proceedings of the International Conference on Society and Information Technologies*, 70–76.

Elendu, C., Amaechi, D. C., Elendu, T. C., Jingwa, K. A., Okoye, O. K., Okah, M. J., Ladele, J. A., Farah, A. H., & Alimi, H. A. (2023). Ethical implications of AI and robotics in healthcare: A review. *Medicine*, *102*(50), e36408.

Fink, M., Maresch, D., & Gartner, J. (2023). Programmed to do good: The categorical imperative as a key to moral behavior of social robots. *Technological Forecasting and Social Change*, *197*, 122923.

Frasca, M., La Torre, D., Pravettoni, G., & Cutica, I. (2024). Explainable and interpretable artificial intelligence in medicine: A systematic bibliometric review. *Discover Artificial Intelligence*, *4*, 12.

Geetha, V., Gomathy, C., Surendar, H., Sanjay, V., & Ramakrishnan, S. (2025). AI in conflict zones: Ethical dilemmas of autonomous warfare. *International Journal of Innovative Research in Science Engineering and Technology*, *14*(3), 112–120.

Huang, C., Zhang, Z., Mao, B., & Yao, X. (2023). An overview of artificial intelligence ethics. *IEEE Transactions on Artificial Intelligence*, *4*(4), 618–639.

Hwang, J. Y. (2024). Ethics of artificial intelligence: Examining moral accountability in autonomous decision-making systems. *World Journal of Advanced Research and Reviews*, *22*(3), 451–463.

Jui, T. D., & Rivas, P. (2024). Fairness issues, current approaches, and challenges in machine learning models. *International Journal of Machine Learning and Cybernetics*, *15*(4), 877–905.

King, A. (2024). Robot wars: Autonomous drone swarms and the battlefield of the future. *Journal of Strategic Studies*, *47*(2), 255–280.

Larsson, S., Liinason, M., Tanqueray, L., & Castellano, G. (2023). Towards a socio-legal robotics: A theoretical framework on norms and adaptive technologies. *International Journal of Social Robotics*, *15*(5), 973–989.

Licardo, J. T., Domjan, M., & Orehovački, T. (2024). Intelligent robotics—A systematic review of emerging technologies and trends. *Electronics*, *13*(3), 567.

Mali, O., & Bhosale, T. (2025). The study of ethics of artificial intelligence and robotics. *International Journal on Science and Technology*, *11*(1), 55–64.

Methnani, L., Chiou, M., Dignum, V., & Theodorou, A. (2024). Who's in charge here? A survey on trustworthy AI in variable autonomy robotic systems. *ACM Computing Surveys*, *56*(9), 1–40.

Nasehi, M. (2025). FPV drone swarms in asymmetric warfare: Tactical innovations and ethical challenges. *Advances*, *4*(2), 99–138.

Pagano, T. P., Loureiro, R. B., Lisboa, F. V., Peixoto, R. M., Guimarães, G. A. S., Cruz, G. O. R., Araujo, M. M., Santos, L. L., Cruz, M. A. S., Oliveira, E. L., Winkler, I., & Nascimento, E. S. (2023). Bias and unfairness in machine learning models: A systematic review on datasets, tools, fairness metrics, and identification and mitigation methods. *Big Data and Cognitive Computing*, *7*(1), 7.

Pham, T. D. (2025). Ethical and legal considerations in healthcare AI: Innovation and policy for safe and fair use. *Royal Society Open Science*, *12*(5), 240408.

Radanliev, P. (2025). AI ethics: Integrating transparency, fairness, and privacy in AI development. *Applied Artificial Intelligence*, *39*(1), 1–24.

Raptis, E. K., Kapoutsis, A. C., & Kosmatopoulos, E. B. (2025). Agentic LLM-based robotic systems for real-world applications: A review on their agenticness and ethics. *Frontiers in Robotics and AI*, *12*, 1523456.

Riesen, E. (2022). The moral case for the development and use of autonomous weapon systems. *Journal of Military Ethics*, *21*(1–2), 46–69.

Sehrawat, V. (2025). Autonomous weapon systems and meaningful human control. *International Journal of Law and Information Technology*, *33*(1), 1–22.

Sholademi, D. B. (2024). Drones and AI in urban security: Monitoring and mitigating threats. *International Research Journal of Modernization in Engineering Technology and Science*, *6*(10), 1205–1213.

Tóth, Z., Caruana, R., Gruber, T., & Loebbecke, C. (2022). The dawn of the AI robots: Towards a new framework of AI robot accountability. *Journal of Business Ethics*, *179*(3), 1–24.

Vouros, G. (2022). Explainable deep reinforcement learning: State of the art and challenges. *ACM Computing Surveys*, *55*(10), 1–38.

Winfield, A. F. T., Swana, M., Ives, J., & Hauert, S. (2025). On the ethical governance of swarm robotic systems in the real world. *Philosophical Transactions of the Royal Society A*, *383*(2273), 20240255.

Ye, Y. (2025). Coexistence challenges in the AI era: Social robots and human networks. *Journal of Global Trends in Social Science*, *4*(2), 77–96.

Ye, Y., You, H., & Du, J. (2023). Improved trust in human-robot collaboration with ChatGPT. *IEEE Access*, *11*, 45732–45747.

14 Project Management for Robotics Systems

14.1 INTRODUCTION

Robotics systems are more than their mechanical design and code (Macenski et al., 2022). Real-world robotics applications often require excellent coordination, planning, and teamwork across individuals with various skills (El Hafi et al., 2022). A robotics project that is likely to succeed from a technical standpoint could be delayed or even fail entirely if it lacks sound project management practices.

Projects often include mechanical engineering, electronics design and prototyping, software engineering, AI behaviour creation, human-machine interface design, and more. Robotics systems must be planned and organised well so that individual engineers and engineers from each discipline can communicate effectively to make sure the robot will work as intended (El Hafi et al., 2022).

Robotics also faces many unknowns. While software projects can also run into unknowns, with robotics there is always a chance that something physically can (and often does) break. Electronics components can fail, sensors can return values you did not expect, and our computers may not integrate with hardware the way we want them to. As such, project planning needs to allow for flexibility and revisiting decisions as you discover new information.

Fortunately, there are project management frameworks we can use to help ourselves deal with complexity (Daraojimba et al., 2024). Agile project management frameworks allow us to break projects down into smaller pieces and focus on completing a series of iterative steps instead of trying to complete the project in one giant step (Appoh et al., 2022).

This chapter will introduce students to project management in robotics. Topics will include Agile methodology, planning in sprints, planning tools we can use, managing risk, and communicating with team members and stakeholders (Edison et al., 2022; Lutsenko, 2023).

14.2 FOUNDATIONS OF PROJECT MANAGEMENT IN ROBOTICS

Project management is defined as how a team plans, organises, and executes their work to ship a product successfully; this includes robot projects as well (El Hafi et al., 2022). Project management is critical in robotics projects as it takes a lot of various complicated processes to design, build, and ship a working robot (Macenski et al., 2022). Project management skills in robotics are harder than managing regular software projects, as robots have hardware you must consider.

Scope, Time, and Resources are three giant pillars that uphold project management. Scope is what you want your robot to do and what you must do to get there.

 DOI: 10.1201/9781003758501-19

Time is when you want things to be done and some milestones you wish to achieve. Resources are the team members, programs, and money you will use to finish your project (Edison et al., 2022). These three constantly affect each other; if you decide to increase your scope, you will most likely need more time and resources.

Robotics projects will generally have a team of engineers and programmers with other specialists like designers. But that will change if you are making a robot for a specific job or industry like nursing, factory production, etc. (Zhang et al., 2025). Teams work best when everyone communicates and knows their role (Strode et al., 2022).

Integration is making sure that everything works together and can perform its duty. All your individual parts may work, but your robot still does not function without integrating (El Hafi et al., 2022). Robotics projects need iteration more than regular software projects. Because you're working with physics, you never know what can go wrong until you build it. Therefore, you will need to keep this in mind when planning.

Documentation will allow you to have everyone on the same page and catch mistakes (García et al., 2022)

14.3 AGILE DEVELOPMENT IN ROBOTICS

Agile development is an iterative project management methodology that prioritises flexibility and adaptation throughout the software development lifecycle (Appoh et al., 2022). It is especially useful for robotics engineering projects because they frequently have elements of uncertainty, shifting requirements, or dependencies on both hardware and software that must be integrated (El Hafi et al., 2022). Planning out every aspect of a robotics system from beginning to end is often impossible.

Some core values of Agile development are iterating on a project in small releases, embracing changing requirements, and frequent communication (Daraojimba et al., 2024). Work is split into small iterations, and team members work to finish small pieces of the system instead of waiting until the deadline to complete everything (Eyieyien et al., 2024). Maintaining a working system always allows for continuous testing and improvement.

Agile development also allows for changing requirements. As you are developing your robotics system, you will likely discover things that you want to change. Having constant or frequent meetings with your team and other stakeholders will help everyone stay on the same page. Developing your hardware and software at the same time will allow you to discover integration problems much earlier.

Agile was originally meant for software development, but its core principles apply to robotics. You may find that you need to have longer iterations to account for physical development. However, you should always have something that can work as a robotics system by the end of an iteration. This allows you to have a working baseline that you can build upon and improve.

The Agile Manifesto values individuals and interactions as well as working software. There are many frameworks that you can use to implement Agile development. One popular framework is called Scrum.

14.4 SCRUM FRAMEWORK

Scrum is an Agile framework that can help robotics teams organise work, collaborate, and respond to challenges during the development process (Daraojimba et al., 2024; Lee & Chen, 2023). This methodology structures work into time-boxed periods known as sprints while defining roles and events that facilitate coordination (Lutsenko, 2023).

There are three primary roles in Scrum. Product Owners serve as representatives of stakeholders. They prioritise tasks and determine which capabilities robots need to work on first (Ghiba, 2022). Scrum Masters are responsible for serving the team by helping people communicate and working through challenges (Appoh et al., 2022). The Development Team consists of everyone who engineers, programs, and designs the robot (Lutsenko, 2023).

The product backlog is where all work is tracked. It's essentially a prioritised list of features or tasks that the robot needs to accomplish (Lee & Chen, 2023). The team pulls items from the product backlog to create a sprint backlog before each sprint, determining what work will be performed during that period.

Sprints are usually one to three weeks long. At the end of each sprint, the team should have something to show for their work. They will develop and test part of the system before reviewing the robot and determining what can be improved (Dzaky & Kurniawan, 2023).

Some of Scrum's events are daily stand-ups where the team provides progress updates, sprint reviews where teams evaluate their output, and retrospectives where people focus on improving teamwork (Appoh et al., 2022; Lee & Chen, 2023).

Scrum works well for robotics projects because it allows teams to embrace iterative processes, test continuously, and adapt to new challenges (Lutsenko, 2023).

14.5 PLANNING TOOLS AND TECHNIQUES

Planning tools are valuable assets for robotics projects, which tend to require interdisciplinary teams working on tightly coupled hardware and software components (El Hafi et al., 2022). The ability to plan helps organise activities across components and teams, keep transparency of progress and coordination efforts (Edison et al., 2022). Structure from planning allows robotics engineers to remain clear on project objectives, stay on schedule, and minimise delays and integration risk (Maravas & Pantouvakis, 2025).

The most common planning tool in many industries is a Gantt chart. These charts are visual representations of a project timeline that are often used for engineering and construction timelines (Maravas & Pantouvakis, 2025). Tasks are added to a timeline with start and end dates. Durations and dependency relationships are then drawn between tasks. This allows teams to identify critical paths, dependencies, and resource demands. For robotics projects, Gantt charts can be applied to higher level phases of design, development, testing, and system integration to help synchronise cross-disciplinary work (El Hafi et al., 2022).

Kanban boards are a popular Agile software development tool that has been leveraged for physical product teams as well. Kanban boards provide visual design cues to help organise tasks into intuitive columns (Edison et al., 2022). Common columns include "To Do," "In Progress," and "Completed" (Strode et al., 2022). Having visual cues as to where each task lies in the development process allows teams to see the workflow and easily identify bottlenecks. From here, it becomes easier to gauge project progress, account for workload, and coordinate.

Task tracking tools allow teams to assign task responsibilities, due dates, and monitor task completion. These tools allow for shared visibility and can generate data that can be used to help forecast timelines and balance workload (Strode et al., 2022). Lastly, teams can benefit from planning around milestone achievements. This can be used to celebrate achievements such as design completion, prototype builds, or integration completions. Milestones are useful because they allow you to track if your team is ahead or behind schedule and identify risks sooner (Maravas & Pantouvakis, 2025).

Planning tools should allow your team to manage complexity without inhibiting flexibility and collaboration (Schlegel et al., 2024).

14.6 RISK ASSESSMENT AND FAILURE PLANNING

Robotics projects are complex and subject to many unknowns. Hardware can fail, software can have bugs, and real-world environments are difficult to predict (Baek & Kröger, 2023; Liu et al., 2022). That is why risk assessment is important: it helps teams identify where and how their project might fail so they can take precautions (Pantano et al., 2022).

Begin by identifying everything that can go wrong. This includes anything from individual components failing to design flaws, delays in the project timeline, running out of resources, or the entire system not working at all (Baek & Kröger, 2023). After you've made your list of possible risks, evaluate each one based on how likely they are to happen and how badly they would affect the project. Even unlikely risks may be worth considering if they would have severe consequences.

After risks have been identified and evaluated, they will need to be prioritised. By organising the risks from most critical to least, teams can ensure that they focus on mitigating the biggest problems first (Pantano et al., 2022). There are many strategies for mitigating risks. Redundancy ensures that if a component fails, there is a backup component to take its place. Rigorous testing can catch bugs before they become serious problems. Contingency plans can be put in place to account for risks that cannot be averted (Baek & Kröger, 2023).

In addition to mitigating risks, it is also important to plan for failures. How will the system behave when something goes wrong? Plans should be made for how to best handle failures when they occur, such as safely shutting down or using fallback behaviours (Hsu et al., 2023).

Risk assessment is an ongoing process. As the project continues, different risks will become more or less likely, and new risks may even arise. This is why it is important to monitor risk mitigation measures as the project progresses (Papoutsakis et al., 2025).

Planning for risks and failures can help teams build safer, more reliable, and resilient systems (Liu et al., 2022; Pantano et al., 2022).

14.7 STAKEHOLDER COMMUNICATION

Robotics projects involve multiple people and disciplines, making communication with stakeholders critical to success (Abayomi et al., 2024; Mojidra et al., 2024). Stakeholders are any individuals or groups who have an interest in the system (Eyeregba et al., 2024). Clients, users, team members, contractors, and outside vendors are examples of stakeholders. Important topics to communicate with stakeholders include updating on progress, managing expectations, seeking feedback, and collaborating.

Progress updates help keep everyone on the same page and allow stakeholders to know what tasks have been completed and what challenges may have arisen. Teams should regularly provide stakeholders with progress reports on the work completed since the last meeting or milestone (Mojidra et al., 2024). This ensures everyone is aware of the system's status and can provide timely feedback.

Expectations should be managed throughout the project. Robotics projects have inherent unknowns that can lead to delays and changing requirements. Informing stakeholders of risks and potential changes early can avoid frustration (Daraojimba et al., 2024). Technical information or jargon should also be avoided so everyone can understand project progress.

Feedback and collaboration can lead to a better overall system. Stakeholders can provide the team with information about requirements or the operating environment. Others may offer suggestions that could save time or money (Appoh et al., 2022; Eyeregba et al., 2024). Collaboration between engineers and domain experts can help create the most well-rounded solution.

There are many methods for communicating with stakeholders, such as meetings, progress reports, presentations, or using project collaboration software. If following an Agile methodology, each sprint review is a time to provide progress updates to stakeholders (Abayomi et al., 2024). Failing to effectively communicate with stakeholders can cause issues down the road when integrating pieces of the project (Mojidra et al., 2024). Problems in communication can lead to parts of the system not working together, mistakes, and costly delays (Daraojimba et al., 2024). Communication is key to maintaining trust with your stakeholders.

14.8 ITERATIVE DEVELOPMENT AND TESTING

Iterative development is developing a system in increments, testing as you go, and iterating until you've reached the goal solution (El Hafi et al., 2022). Rather than developing a product or solution once through, multiple iterations of development/testing/fixing occur (Colle et al., 2025; Leone et al., 2025).

Prototyping starts the process by creating a preliminary model of something to test some high-level ideas without necessarily having a completely functioning product (Edwards et al., 2025; Gebellí & Ros, 2025). Find as many unknown unknowns as you can early on to help limit your exposure to risk (Nassiuma et al., 2026).

Testing can be done in simulation, in a lab environment, or out in the real world. Things will inevitably break or not behave how they should once tested (Pandy et al., 2025; Zhang et al., 2025).

Finally, teams fix everything that was broken and tweak everything else to be better than before. This could include fixing algorithms, changing a part's shape to fit together better, or changing the way the system behaves under certain conditions. One testing and subsequent refinement is 1 iteration (Pandy et al., 2025).

Make sure that your components are always able to communicate with each other as you develop them (continuous integration) (El Hafi et al., 2022). Getting things out for testing early also allows you to receive feedback quicker and avoid expensive failures near a project's deadline.

Robotic systems are also extremely difficult to predict how they will behave until they are built and tested. Iteration allows you to discover these unknowns and reiterate your design to handle them. Repeat this process again until you've reached your goal. This aligns with the Agile philosophy of development (Goyal, 2023)

14.9 PRACTICAL INSIGHT: SPRINT PLANNING FOR A ROBOTICS PROJECT

Sprint planning can be used for planning what work should be done for a given timebox, usually one or two weeks (Lieberum et al., 2022). Firstly, the team can decide what work needs to be done for the robot as part of sprint planning. Usually, robotics projects have many tasks related to hardware, software, and testing; sprint planning can help with organising these tasks (Abadi et al., 2025).

To start a sprint, the team needs to decide what items from the product backlog will be completed during the sprint (Daraojimba et al., 2024). The product backlog contains all the tasks/features needed to complete the product (Daraojimba et al., 2024). These items are chosen based on priority and are added to the sprint backlog (Abadi et al., 2025). Ideally, the sprint backlog should consist of items that can be completed within a sprint and that provide a valuable upgrade to the system (Kula et al., 2024).

After that, each task from the sprint backlog is divided into small actionable steps that can be assigned to team members based on their role/responsibilities (Abadi et al., 2025; Edison et al., 2022). For instance, one member can be assigned the task of integrating sensors while another works on control algorithms/testing.

The team should come up with a goal that will be worked towards during the Sprint. For instance, the robot should be able to perform basic navigation by the end of Sprint or reach X% accuracy in object detection. The team's progress can be measured by whether they reach this goal or not (Kula et al., 2024).

Listing 14.1: Example sprint plan (conceptual structure)

```
sprint_goal = "Implement obstacle avoidance system"

sprint_tasks = [
    "Integrate LIDAR sensor",
    "Develop obstacle detection algorithm",
    "Test sensor data accuracy",
    "Implement avoidance behaviour",
    "Run simulation tests"
]
```

```
team_assignments = {
    "Engineer A": "Sensor integration",
    "Engineer B": "Algorithm development",
    "Engineer C": "Testing and validation"
}

for member, task in team_assignments.items():
    print(f"{member} is responsible for {task}")
```

During a sprint, the team may measure progress via a Kanban board or daily stand up. At the end of the sprint, the team reflects on what was accomplished and what can be improved during the next sprint. Sprint planning allows your work to be focused, organised, and goal oriented. Splitting up development into these iterations allows you to quickly overcome obstacles, keep moving forward and complete your entire robot.

14.10 CHALLENGES IN ROBOTICS PROJECT MANAGEMENT

Robotics projects can be difficult for several reasons:

Complexity: Robotics involves multiple disciplines like mechanics, electronics, and computer science. They're physical systems with real-world interactions, which adds unpredictability (El Hafi et al., 2022).

Real-world environment: Robotics projects often involve testing in uncontrolled, real-world environments, which can lead to unexpected challenges and issues (El Hafi et al., 2022).

Iterative and interdependent development: Teams may use Agile or Scrum development to divide into two-week sprints (Lawong & Akanfe, 2024). However, some work can get blocked by dependencies or bugs in other areas (Keshvarparast et al., 2023).

Timeline slippage: Debugging hardware and software issues can cause projects to fall behind schedule (Macenski et al., 2022).

Integration bugs: Parts of the robot may function independently, such as sensors, controllers, and code modules, but cause errors when integrated together. Solving communication issues between subsystems can take many iterations to debug (Macenski et al., 2022).

Coordination challenges: Collaborative robotics teams that aren't attuned to one another can experience slowdowns if people on different parts of the team aren't communicating well. It can be especially difficult for members of multidisciplinary teams to understand each other (Lawong & Akanfe, 2024).

Scope creep: The requirements of a project may grow to improve the robot's performance. However, if the scope of a project continues growing, it can become overwhelming given time and resource constraints (Salimimoghadam et al., 2025).

Unknown unknowns: Variables in the real world can cause a robotics system to behave in ways that you wouldn't have found during testing.

Resource limitations: You may have restrictions on time, money, or development resources (Keshvarparast et al., 2023).

Some of these challenges can be avoided or mitigated through foresight and planning (El Hafi et al., 2022). However, others require you to think on your feet and adapt (Macenski et al., 2022).

14.11 ETHICAL AND PROFESSIONAL CONSIDERATIONS

Robotics projects must consider important ethics and professional concerns during development: Responsibility, Safety, Professionalism, Teamwork, and Value-sensitive design. Ignoring these will cause teams to create systems unfit for real-world use (Winfield et al., 2025).

Responsibility refers to taking ownership of the systems teams create. Did the robots behave as expected? Have teams created unexpected risks? Engineers must validate their systems and ensure robots are thoroughly tested and safe before use (Winfield et al., 2025).

Safety matters because robots can cause injuries to humans and damage to the environment. Responsible teams consider how a robot could cause harm, then design appropriate safety features (Panagou et al., 2023). This includes fail-safe behaviours and emergency stop mechanisms.

Professionalism ensures systems are engineered to high standards. Teams must create quality code and documentation, raise issues as soon as they arise, and communicate honestly and transparently (Tóth & Blut, 2024).

Teamwork ethics include treating each team member with respect and crediting teammates for their work. Following good teamwork ethics will help teams develop projects responsibly (Panagou et al., 2023).

Teams should also consider the social impact of their designs. Will the system have positive impacts on its users? Value-sensitive design tries to anticipate every effect a new technology has on society, allowing teams to address issues early on (Söderlund, 2023). These include privacy violations, accessibility, and malicious usage of technology.

Keeping ethics in mind throughout the design process will allow for the creation of safer, highly trustworthy systems (Johnston, 2022).

14.12 CONCLUSION

The chapter provided an overview of project management fundamentals. It emphasised the significance of project management practices for building successful robotic systems. Developing a robot was presented as an engineering and organisational accomplishment that depends on planning, teamwork, and communication.

Agile methodologies such as Scrum were described as helpful for dealing with complex and uncertain projects. Gantt charts, Kanban boards, and sprint planning were introduced as practical tools for organising work and keeping track of progress. Risk management, failure modes and effects analysis (FMEA), and stakeholder communication were some of the concepts that showed how teams could plan for and react to potential setbacks.

Iterative development and testing were recommended as ways to cope with the complexity and uncertainty of robotics systems. In addition to explaining project management terminology and concepts, the chapter also covered common difficulties encountered while building robotic systems. Including delays, technical glitches, budget restrictions, etc. The significance of adaptability and flexibility was highlighted. Ethical and professional issues were also mentioned to remind teams to build safe and reliable systems that do not negatively impact society.

The chapter made it clear that successfully developing robotics systems requires more than just technical knowledge. Teams must also use solid project management techniques to the best of their abilities.

14.13 DISCUSSION QUESTIONS

1. Why is Agile development well suited to robotics projects?
2. Robotic projects face unique management challenges. What are they?
3. How do project planning tools help with coordination?
4. What makes risk assessment critical for robotics developers?
5. Delays and surprises are inevitable. How can we recover from them as a team?

14.14 EXERCISES

Exercise 1: Sprint Planning

Plan a two-week sprint with related tasks, roles, and goals.

Exercise 2: Kanban Board

Design a Kanban board for task management.

Exercise 3: Risk Assessment

Identify potential project risks and how to mitigate them.

Exercise 4: Reflection

If a robotics project falls behind schedule, what happens? How can we course correct as a team?

REFERENCES

Abadi, A., Abadi, C., & Abadi, M. (2025). Intelligent decision making and knowledge management system for agile project management in Industry 4.0 context. *Statistics, Optimization & Information Computing*, *13*(1), 1–24.

Abayomi, A. A., Ogeawuchi, J. C., Gbenle, T. P., Agboola, O. A., & Uzoka, A. C. (2024). Advances in project stakeholder communication and transparency using cloud collaboration platforms. *International Journal of Scientific Research in Science and Technology*, *11*(5), 1–15.

Appoh, M., Frempong, D., Akinboboye, O., Okoli, I., Afrihyia, E., Umar, M. O., Umana, A. U., & Omolayo, O. (2022). Agile-based project management strategies for enhancing collaboration in cross-functional software development teams. *Journal of Frontiers in Multidisciplinary Research*, *1*(2), 45–62.

Baek, W. -J., & Kröger, T. (2023). Safety evaluation of robot systems via uncertainty quantification. In 2023 IEEE International Conference on Robotics and Automation (ICRA) (pp. 3989–3995). IEEE.

Colle, A., Donaldson, K., & Dragone, M. (2025). Co-design methodology for rapid prototyping of modular robots in care settings. *Frontiers in Robotics and AI*, *12*, 1–22.

Daraojimba, E. C., Nwasike, C. N., Adegbite, A. O., Ezeigweneme, C. A., & Gidiagba, J. O. (2024). Comprehensive review of agile methodologies in project management. *Computer Science & IT Research Journal*, *5*(1), 190–218.

Dzaky, F. A., & Kurniawan, D. (2023). Implementation of Scrum framework Agile method to develop integrated asset management information system at Universitas Diponegoro inventory module. *Jurnal Masyarakat Informatika*, *5*(2), 101–110.

Edison, H., Wang, X., & Conboy, K. (2022). Comparing methods for large-scale Agile software development: A systematic literature review. *IEEE Transactions on Software Engineering*, *48*(8), 2964–2993.

Edwards, A., Saudagar, S., Fatima, Z., Tanveer, M. H., Desa, H., & Voicu, R. (2025). Development of a rapidly prototyped automated robotic manufacturing system. In 2025 9th International Conference on Man-Machine Systems (ICoMMS) (pp. 1–6). IEEE.

El Hafi, L., Garcia Ricardez, G. A., von Drigalski, F., Inoue, Y., Yamamoto, M., & Yamamoto, T. (2022). Software development environment for collaborative research workflow in robotic system integration. *Advanced Robotics*, *36*(11–12), 595–610.

Eyeregba, M. E., Mokogwu, C., Ewim, S. E., & Olorunyomi, T. D. (2024). Developing a framework for business analysts to bridge stakeholder and technical team gaps: Enhancing collaboration and project success. *International Journal of Applied Research in Social Sciences*, *7*(4), 90–115.

Eyieyien, O. G., Idemudia, C., Paul, P. O., & Ijomah, T. I. (2024). Advancements in project management methodologies: Integrating agile and waterfall approaches for optimal outcomes. *Engineering Science & Technology Journal*, *6*(2), 55–71.

Gebellí, F., & Ros, R. (2025). An in-situ participatory approach for assistive robots: Methodology and implementation in a healthcare setting. *Frontiers in Robotics and AI*, *12*, 1–23.

Ghiba, A. C. (2022). The product owner role in a contemporary Agile team. Annals of Dunarea de Jos University of Galati. *Fascicle I. Economics and Applied Informatics*, *28*(1), 35–43.

Goyal, A. (2023). Integrating IoT and Agile methodologies for smarter engineering solutions. *International Journal of Science and Research Archive*, *8*(1), 123–132.

Hsu, K.-C., Hu, H., & Fisac, J. F. (2023). The safety filter: A unified view of safety-critical control in autonomous systems. *Annual Review of Control, Robotics, and Autonomous Systems*, *7*, 1–27.

Johnston, C. (2022). Ethical design and use of robotic care of the elderly. *Journal of Bioethical Inquiry*, *19*(1), 27–39.

Keshvarparast, A., Battini, D., Battaia, O., & Pirayesh, A. (2023). Collaborative robots in manufacturing and assembly systems: Literature review and future research agenda. *Journal of Intelligent Manufacturing*, *34*(8), 3439–3470.

Kula, E., van Deursen, A., & Gousios, G. (2024). Context-aware automated sprint plan generation for agile software development. 39th IEEE/ACM International Conference on Automated Software Engineering (ASE), 1–12.

Lawong, D. A., & Akanfe, O. (2024). Overcoming team challenges in project management: The Scrum framework. *Organizational Dynamics*, *53*(2), 100960.

Lee, W.-T., & Chen, C. -H. (2023). Agile software development and reuse approach with Scrum and software product line engineering. *Electronics*, *12*(15), 3210.

Leone, S., Lago, F., Pîslă, D., & Carbone, G. (2025). A systematic approach for robotic system development. *Technologies*, *13*(4), 85.

Lieberum, T., Schiffels, S., & Kolisch, R. (2022). Should we all work in sprints? How agile project management improves performance. *Manufacturing & Service Operations Management*, *24*(3), 1453–1472.

Liu, J., Wang, D., Lin, Q., & Deng, M. (2022). Risk assessment based on FMEA combining DEA and cloud model: A case application in robot-assisted rehabilitation. *Expert Systems with Applications*, *200*, 117028.

Lutsenko, G. (2023). Team-projects in learning Scrum methodology for engineering students. *Engineering and Educational Technologies*, *11*(1), 45–59.

Macenski, S., Foote, T., Gerkey, B. P., Lalancette, C., & Woodall, W. (2022). Robot operating system 2: Design, architecture, and uses in the wild. *Science Robotics*, *7*(66), eabm6074.

Maravas, A., & Pantouvakis, J. -P. (2025). Schedula Anima: Dynamic visualization of Gantt charts and resource usage graphs in project scheduling. *Buildings*, *15*(3), 393.

Mojidra, M., Pitroda, J. R., Shah, J. P., & Gohil, J. (2024). The impact of stakeholder communication and coordination on project outcome. *International Journal of Scientific Research in Engineering and Management*, *8*(11), 150–164.

Nassiuma, I., Hubbard, E.-M., & Goh, Y. M. (2026). High-fidelity simulation and digital twin framework for real-time adaptive human-robot collaboration. In AHFE 2026 International Conference (pp. 1–12). Springer.

Panagou, S., Neumann, W. P., & Fruggiero, F. (2023). A scoping review of human-robot interaction research towards Industry 5.0 human-centric workplaces. *International Journal of Production Research*, *61*(16), 5560–5583.

Pandy, G., Pugazhenthi, V. J., Jeyarajan, B., & Murugan, A. (2025). Advancing robotics testing: A novel framework for adaptive and scalable evaluation. In 2025 17th International Conference on Computer and Automation Engineering (ICCAE) (pp. 1–7). IEEE.

Pantano, M., Pavlovskyi, Y. P., Schulenburg, E., Traganos, K., Ahmadi, S., Regulin, D., Lee, D., & Saenz, J. F. (2022). Novel approach using risk analysis component to continuously update collaborative robotics applications in the smart, connected factory model. *Applied Sciences*, *12*(11), 5615.

Papoutsakis, M., Hatzivasilis, G., Michalodimitrakis, E., Ioannidis, S., Michael, M. K., Savva, A. D., Nikolaou, P., Stokkou, E., & Bozdemir, G. (2025). SESAME: Automated security assessment of robots and modern multi-robot systems. *Electronics*, *14*(5), 1045.

Salimimoghadam, S., Ghanbaripour, A., Tumpa, R. J., Rahimi, A. K., Golmoradi, M., Rashidian, S., & Skitmore, M. (2025). The rise of artificial intelligence in project management: A systematic literature review of current opportunities, enablers, and barriers. *Buildings*, *15*(4), 905.

Schlegel, D., Rosenberg, B., Fundanovic, O., & Kraus, P. (2024). How to conduct successful business process automation projects? An analysis of key factors in the context of robotic process automation. *Business Process Management Journal.*

Söderlund, M. (2023). Service robots and artificial morality: An examination of robot behavior that violates human privacy. *Journal of Service Theory and Practice*, *33*(6/7), 457–479.

Strode, D., Dingsøyr, T., & Lindsjørn, Y. (2022). A teamwork effectiveness model for Agile software development. *Empirical Software Engineering*, *27*, 37.

Tóth, Z., & Blut, M. (2024). Ethical compass: The need for corporate digital responsibility in the use of AI in financial services. *Organizational Dynamics*, *53*(1), 100947.

Winfield, A. F. T., Swana, M., Ives, J., & Hauert, S. (2025). On the ethical governance of swarm robotic systems in the real world. *Philosophical Transactions A*, *383*(2281), 20240260.

Zhang, G., Luo, X., Zhang, L., Li, W., Wang, W., & Li, Q. (2025). A framework of indicators for assessing team performance of human–robot collaboration in construction projects. *Buildings*, *15*(8), 2043.

Part VI

The Future of Intelligence

15 Innovation, Futures, and Hybrid Intelligence

15.1 INTRODUCTION

Robotics and artificial intelligence (AI) are rapidly advancing fields that continue to shape our world in unimaginable ways (Licardo et al., 2024). Systems are becoming increasingly capable of adapting to new situations and learning from their experiences, leading to a world where robots can interact with humans in ways never seen before (Licardo et al., 2024). Human–machine collaboration will become much more prevalent in the future of robotics, moving away from simply automating tasks to instead collaborating and working together to achieve a common goal. We are now living in the age of hybrid intelligence.

Many topics have been discussed during this book, such as perception, control, learning, and interaction. Now these topics will come together to create smart systems that can adapt to their environment, continuously learn, and interact with humans to complete tasks in any feasible environment (Cohen et al., 2025). The robots of the future will be more adaptable than ever before and will be able to react to their surroundings and context in real-time (Urrea & Kern, 2025).

Technological advancements and societal needs will drive innovation in the field of robotics (Haidegger et al., 2023). Machine learning, new sensors, and new materials will allow for new robots to be created, while global issues will allow us to apply these technologies to solve problems and help people.

This chapter aims to discuss what the future of robotics will look like. We will discuss some emerging technologies and define hybrid intelligence. We will also reflect on how robotics will affect society, your careers, and the world (Haidegger et al., 2023). Lastly, we will have you think about the future to help you learn about what you can do to help shape that future (Licardo et al., 2024).

15.2 FOUNDATIONS OF INNOVATION IN ROBOTICS

Robotics can advance through a variety of means including new technology, identifying problems, iteration, collaboration between various fields of technology, and new trends or disruptions (Licardo et al., 2024; Urrea & Kern, 2025). This is why robotics can advance so quickly.

Robotics advances through technology because the more advanced AI, machine learning, sensor technology, materials technology, and battery technology become, the more robots will become able to do things they couldn't do before. Sensors can allow robots to better map their environment, while new materials can allow for soft robotics. Machine learning allows robots to take information and act differently if exposed to the same situation in the future.

DOI: 10.1201/9781003758501-21

Robotics also advances when there is a problem that needs to be addressed. Robots are used for an abundance of things to help people with medical needs, deliver goods, and help people with daily tasks. The global demand for robotics systems helps drive what research is being done to create the systems that we want to see.

Iteration allows for robots to be developed through multiple prototypes until the product works how we want it to. Rather than having a single person dream up a robotics system and have it built immediately, it will usually go through many cycles of testing and editing.

Collaboration between other fields of tech can also propel robotics. Robotics itself is a mixture of engineering and computer science and can include psychology and knowledge of biology. Someone from a different field may be able to help solve a problem in robotics.

Lastly, new trends and disruptions can propel it forward. As soon as we develop new technology, we can find a way to integrate it into robotics. Something as simple as AI has changed the game for robotics.

15.3 HYBRID INTELLIGENCE SYSTEMS

Hybrid intelligence is a model of human–machine interaction where both parties work together and leverage each other's strengths (Järvelä et al., 2025). This is increasingly applied to robotics and represents a change of paradigm from automation, which replaces human work, to augmentation, which assists it (Puerta-Beldarrain et al., 2024).

As experts above have stated, humans and robots have very different strengths they excel at. Humans are better at creativity, complex social understanding, and ethics among other things (Järvelä et al., 2025). Robots are very good at physical actions, consistently performing tasks they have been programmed to perform, and processing large amounts of information rapidly and without getting tired (Cohen et al., 2025). Hybrid intelligence systems allow both humans and robots to work to their strengths and complement each other (Puerta-Beldarrain et al., 2024).

Systems using hybrid intelligence can come in many forms. One could have an operator supervise a robot completing a task, ready to take over should anything go wrong. Or a robot may assist humans with parts of their job – providing recommendations to human decision makers, processing information to analyse data, or taking over routine parts of a task to let humans focus on higher level decisions. The robot may still be operating under the supervision of the human, or it may carry out decisions autonomously (Puerta-Beldarrain et al., 2024).

These systems can be particularly useful in complex or unstructured environments – such as hospitals, disaster response or high-tech manufacturing – where it might be difficult to program a robot to handle every eventuality. But it still benefits from having a human in the loop to provide judgement and flexibility (Hopko et al., 2022).

Hybrid intelligence can also come in the form of shared control, where neither the human nor the robot is in full control of the process and both influence decisions to some degree (Yin et al., 2026). As you might imagine, this requires seamless coordination and communication between human and robot (Borghoff et al., 2025).

Hybrid intelligence also comes with challenges. Trust, responsibility and responsibility shift, user experience and interface design can all play a part in the success or failure of a human–machine team (Hopko et al., 2022). Ultimately, how these two agents are combined will influence the capability of the system (Cohen et al., 2025).

Hybrid intelligence is an exciting field and an important area to consider for the future of robotics. Machines never stop learning and will only continue to become better integrated in our lives. Hybrid intelligence allows us to capitalise on these strengths of this new technology while still maintaining human control where it is needed.

15.4 CASE STUDY: ASIMO AND OpenAI

ASIMO and OpenAI can be seen as two sides of one coin since each focuses on either physical or cognitive technology. However, together they demonstrate the concept of hybrid intelligence (Järvelä et al., 2025).

ASIMO emphasises embodied intelligence. Honda's humanoid robot could walk, running on two legs and communicating with humans through voice commands. ASIMO was made to work among humans, so its primary purpose revolved around moving, balance, and interacting with the environment, which made it integral to service robots and assistive technology (Göngör & Tutsoy, 2024).

OpenAI is based around large language models, which attempt cognitive intelligence. Large language models process and output human language and allow machines to reason about different topics and help them complete tasks. However, large language models are typically not paired with a physical bot that can move around or interact with their environment (Yu et al., 2023). Instead, these bots help us become more intelligent.

AI is becoming embodied just as robots are becoming more intelligent. There has been a rise in bots that can sense their environment, learn from data, and complete actions. Robots of the past were able to complete actions. Modern robots can now sense their environment and learn from data. The robots of the future will think.

Combining the two fields allows us to create large language models that can physically act on our planet while also creating bots that can learn from physical interaction.

15.5 SOFT ROBOTICS AND EMERGING TECHNOLOGIES

Soft robotics is a field of robotics that utilises compliant materials, meaning robots made of soft materials instead of traditional rigid materials. Robots made with soft components can be advantageous for various reasons, such as operating safely in unstructured environments and more naturally interacting with humans and the real world (Zhou et al., 2024). Soft robotics design commonly draws inspiration from real-life biological organisms such as octopus tentacles or muscles to allow for flexibility in movement and versatility in controlling their shapes (Sarker et al., 2024).

The major benefit of soft robotics is its adaptability. Soft robots made from elastic materials excel at gripping and manipulating objects with unusual or irregular surfaces. Soft robotics can deform to traverse unstructured terrain and confined spaces (Yin et al., 2024). This property allows soft robots to handle objects that should not be subjected to rigid robotics. Also, because soft robotics is more compliant, they are safer for human–robot interaction (HRI) (Giordano et al., 2023).

The progress in soft robotics is propelled by advances in material design, sensing, and AI. Materials with the ability to react to their surroundings, such as pressure or heat, are called smart materials. Advances in sensor technology allow for finely tuned responses to feedback about the robot's environment. Machine learning allows these robots to learn from their surroundings and change behaviours accordingly (Qu et al., 2024).

Applications of soft robotics include use in hospitals, factories, and homes. Some of these applications can include robots that aid in surgery or are wearable support devices. Other benefits of soft robotics include less energy usage and safer interactions. Moving forward, soft robotics will continue to help design collaborative robots that resemble natural movement.

15.6 BRAIN–COMPUTER INTERFACES (BCIs)

Brain–computer interfaces (BCIs) are systems that allow communication between the human brain and an external device, such as a computer (Edelman et al., 2024). BCIs allow people to control robots using their thoughts, instead of using traditional input devices like keyboards or joysticks (Dong et al., 2023).

BCIs work by reading brain signals and translating them into commands that a robot can understand. The interface often involves placing electrodes on the scalp to detect electrical activity in the brain. The signals are then processed by algorithms that can interpret the user's intention and convert them into commands for the robotic system (Edelman et al., 2024). For instance, this technology can allow users to move a robotic arm or wheelchair using their thoughts.

BCIs can be used for assistive technology purposes, allowing users with physical disabilities to interact with robots. For example, a person with paralysis could use a BCI to control a robotic arm and manipulate objects in their environment (Qu et al., 2022). This technology can help people with disabilities become more independent and engaged with the world around them (Karas et al., 2023).

BCIs can also be used in what's known as hybrid intelligence systems, where the intention signal is directly obtained from the human brain and translated into actions by a robot. This allows for a more seamless integration of human intention and robot action (Dong et al., 2023; Edelman et al., 2024).

There are some challenges associated with using BCIs in robotics. Interpreting brain signals can be difficult, as they are often noisy and ambiguous (Dong et al., 2023). Additionally, training users to control a BCI effectively can take time and effort. There are also ethical concerns around privacy and security, as BCIs involve collecting sensitive brain data from users. Overall, while BCIs present some challenges, they offer a direct way of linking human intention to robotic action and represent an exciting area of research in human–machine interaction (Edelman et al., 2024). Continued development of this technology could lead to more intuitive and effective ways for people to interact with robots.

15.7 SOCIOECONOMIC AND ENVIRONMENTAL DRIVERS OF INNOVATION

Robotics and AI innovation is influenced not only by advances in technology but also by market demand, economic factors, social needs, environmental concerns, and political will (Haidegger et al., 2023).

There is economic demand for robots and AI to increase efficiency, lower costs, and perform tasks that humans cannot or will not do. As a result, robots have been widely adopted in various industries such as manufacturing, logistics, and customer service. The demand for automation can also create fear of job losses and challenges around transitioning workers into new roles.

Society's needs and challenges also influence robotics innovation. As our society faces issues such as an ageing population, increased demand for healthcare and assistive technologies, and urbanisation, robotics technology has been developed to meet those needs. This includes the creation of robots to assist with elder care, exoskeletons to help workers with physical tasks, and self-driving cars and aircraft (Asgharian et al., 2022).

Environmental concerns can also drive innovation in robotics. Robots can be used in applications such as precision agriculture, wildlife conservation, and recycling to help improve efficiency and minimise waste. Additionally, robots can be used in monitoring and data collection to study and track environmental changes (Giordano et al., 2023). Many of the world's pressing concerns, such as climate change, also create a need for energy-efficient and environmentally friendly robots (Bolón-Canedo et al., 2024).

The government can influence robotics innovation through policy and regulation. They can either encourage the development of robotics through funding and tax incentives or slow it down through restrictions and regulations (Haidegger et al., 2023).

The general feeling of the public can influence how robots are accepted. If people accept and trust the technology, it will be more widely adopted. If not, the technology will face many barriers to entry.

15.8 CAREERS AND FUTURE PATHWAYS

Robotics will provide a wide array of career opportunities in various industries (Licardo et al., 2024). This is due to robotics being extremely interdisciplinary (Licardo et al., 2024). Branching from technical fields such as engineering, computer science, data science, and design, individuals can enter many career paths within robotics (Martínez Elizalde et al., 2025).

Robotics engineers oversee creating robots by designing, building, and testing robotics systems (Licardo et al., 2024). They typically work more with hardware such as robotics components, sensors, actuators, and control systems (Licardo et al., 2024). Software engineers program robots by focusing on writing code, developing algorithms, and architecting systems to make robots behave intelligently (Martínez Elizalde et al., 2025).

AI and machine learning engineers write the models that allow robots to learn from data and act without supervision (Kim et al., 2024). HRI and UX designers focus on bridging the gap between robots and humans in intuitive, responsible, and ethical ways (Mårell-Olsson et al., 2025).

Careers in ethics, policy, and regulation will become more prevalent as robots become more autonomous and we need to ensure these systems do not harm our society and can be safely interacted with (Etemad-Sajadi et al., 2022).

Robotics also offers career opportunities in research, academia, management, consulting, sales, and entrepreneurship (Licardo et al., 2024). Many of these careers will require continuous education, upskilling, and cross-skilling to stay on top of your field, in addition to having creativity and strong technical skills.

15.9 PRACTICAL INSIGHT: SCENARIO PLANNING FOR 2035

Scenario planning exercises entail "envisioning how current trends may unfold over time" (Finch et al., 2023). The purpose is to paint a realistic picture of potential futures. Scenario planning has been used in robotics education to help students understand how future workplaces might operate. For example, students consider how hybrid intelligence, AI, and automation will affect jobs and society over the next several decades (Järvelä et al., 2025).

Students start by selecting several important trends in robotics, AI, or automation. For example: growing AI, increasing automation of tasks, shifting workforce demands, and human–machine interaction developments (Weidemann et al., 2023). These trends will be the foundation for scenario development (Finch et al., 2023).

Next, learners will pick a scenario context. In robotics education, a scenario context may be the workplace in 2035. Students can select any type of workplace of interest, such as hospitals, factories, offices, delivery trucks, service vehicles, and smart cities. Educators should encourage learners to think broadly about where robots may be when imagining the context of a scenario.

Students then identify features of this future context. In our example, learners may consider how robots will work collaboratively with humans in the workplace. What jobs are automated, and what duties will humans be responsible for? (Weidemann et al., 2023).

Educators may provide students with guidance for considering specific elements of these future scenarios. For example, one aspect is clearly defining people's roles and how they interact with machines. In a hybrid-intelligent workplace of the future, most people will work alongside AI (Verma & Singh, 2024). However, people will handle decision-making and creative tasks. Robots will take on mundane tasks and calculations. Human–machine interactions also may look different. In the future, people may interact with machines through voice, hand gestures, BCIs, etc. (Lv et al., 2022).

Students can develop scenarios using the following outline:

Listing 15.1: Scenario planning structure

```
scenario = {
    "year": 2035,
    "environment": "Smart logistics warehouse",
    "technologies": ["AI systems", "Autonomous robots", "BCI
interfaces"],
```

```
    "human_roles": ["Supervision", "Decision-making", "System
design"],
    "robot_roles": ["Transport goods", "Optimise routes",
"Monitor inventory"],
    "interaction": "Humans and robots collaborate through shared
control systems"
}

for key, value in scenario.items():
    print(f"{key}: {value}")
```

Students can further develop the scenario by brainstorming potential obstacles/challenges, risks, and ethical dilemmas. Will trust issues arise? How do they recover from failures? How do their roles shift/change? Scenario planning allows students to think creatively while still rooting their concepts in the real-world applications of today. Students begin to realise that they aren't just dreaming about the future of robotics – they can help create it.

15.10 CHALLENGES IN FUTURE ROBOTICS

Robotics challenges can range from technical issues to ethical concerns, social impacts, and environmental considerations.

Uncertainty about the future. Robotics and artificial intelligence's future trajectory are unknown, including what systems will be developed and how they will be used. Planning for the long-term effects of robotics can be difficult because of this (Zhao & Gómez Fariñas, 2022).

Ethical and moral risks increase with increased system autonomy. Robots programmed to make choices without human intervention might harm individuals if their morals are not aligned with society's. Ensuring that autonomous systems behave ethically can also be difficult as they become more complicated (Hwang, 2024).

Quick technological advancements may also be problematic. Society may have difficulty adapting to robotics as it advances because humans and organisations would need to keep learning new systems and technologies (Shengelia, 2025).

Systems may not always be accepted by society. Robots must be created with public sentiments in mind to be widely accepted, as people can be worried about automation. This includes worries about privacy, safety, and job displacement (Gerlich, 2023).

There is a risk of increasing economic inequality. While robotics has the potential to raise productivity, only those who can afford to invest in robotics will initially benefit. This can cause the rich to become richer while further impoverishing the poor who cannot afford robots (Shengelia, 2025).

Robotics may have a significant environmental impact if it is not designed sustainably. Robotics creation and operation require resources and energy, so they must be built in ways that reduce environmental damage (Bolón-Canedo et al., 2024).

15.11 REFLECTION: THE ROLE OF THE TECHNOLOGIST

There are several things that come with being a technologist. When you design, engineer, and develop systems that literally help build the future, you can help sculpt how technology influences society. Responsibility is such a large part of robotics and AI. Engineers and designers must be smart about how we reflect on our values, our choices, and how they'll affect our futures (Stahl & Eke, 2024).

Responsibility: How will your system be used and by whom? What is its purpose? How will your system affect people and the planet? Designing ethical and responsible systems that use robotics and AI requires thinking about how your technology is safe, how it treats different groups of people fairly, and how it aligns with moral rules or values. We also must think about how our systems can be abused or used negatively and prevent unintended consequences (Radanliev, 2025).

Identity and purpose. What sort of work do you want to be doing? Some people care more about solving technological problems. Others want to design systems with a social impact. There are technologists who care about sustainability or creating accessible technology for as many people as possible. Figuring out what your why is can help shape what you work on and where you spend your time (Palenski et al., 2024).

Lifelong learner: The world of robotics is constantly changing and growing. There's always going to be things to learn, whether it be new programming languages, wonderful applications of new technology, or fresh ethics dilemmas to evaluate (Kamalov et al., 2023). Part of your job as a technologist is making sure you learn as much as you can about your field and about how technology interacts with people and society.

Teacher or communicator: Whether you have to explain to others how your system works, speak at a conference about your experiences, or write on your blog. Communicating your thoughts and ideas about technology helps create better discourse about how technology is made and how it's used by the public.

Role model: Robots and the technology you create today could be around for decades into the future. If you take care and think about the longer "role" you have when creating technology, you can help improve how technologies are created in the future. Think big and spend time crafting your ethical responsibility and mentality towards technology.

This reflection asks students to consider their role in the future of robotics. Integrating technical expertise with ethical consciousness and a sense of purpose can guide technologists in designing systems that promote both advancement and positive societal impact.

15.12 CONCLUSION

The chapter provided valuable insights into the future of robotics. It focused on how robotics is shifting from standalone systems to collaborative, adaptable, and human-centric technologies, emphasising the importance of hybrid intelligence. Students will learn how technologies such as soft robotics and BCIs are paving the way for future developments.

The chapter showcased examples of current technologies that highlight how boundaries are being pushed for new ways of thinking about human and machine interaction. The comparison of ASIMO and OpenAI highlighted the potential for merging physical and cognitive systems in the future.

Environmental and socioeconomic factors were identified as potential driving forces for innovation. The career opportunities section allows students to explore the many career paths available to them if they wish to pursue a career in robotics.

Uncertainty, ethics risks, and societal impact were identified as challenges that we may face. Reflecting on our role as technologists can help us better understand how to approach these challenges.

In conclusion, the future of robotics is not only about advanced technology but also about people and how we can work together to responsibly innovate for the betterment of society. Students can shape the future of robotics and AI by understanding the key takeaways from this chapter.

15.13 DISCUSSION QUESTIONS

1. What is hybrid intelligence? Why will it be essential for the future of robotics?
2. How will new technologies, such as BCIs, alter human–robot interaction?
3. What factors will influence robotics innovation?
4. In what ways will robotics influence careers and the future of work?
5. What responsibilities do technologists have to the future?

15.14 EXERCISES

Exercise 1: Designing Your Future

Imagine that it's the year 2035. Describe a day in the life of working in a hybrid-intelligent workplace.

Exercise 2: Robotics Career Path

Find a robotics or AI career that appeals to you. Describe the educational and professional requirements.

Exercise 3: Robotics Innovation

Find a current news article about robotics innovation. How will it impact society?

Exercise 4: Reflection

What type of technologist do you want to be? What are your core values that will guide your work?

REFERENCES

Asgharian, P., Panchea, A., & Ferland, F. (2022). A review on the use of mobile service robots in elderly care. *Robotics*, *11*(6), 131.

Bolón-Canedo, V., Morán-Fernández, L., Cancela, B., & Alonso-Betanzos, A. (2024). A review of green artificial intelligence: Towards a more sustainable future. *Neurocomputing*, *574*, 127015.

Borghoff, U. M., Bottoni, P., & Pareschi, R. (2025). Human-artificial interaction in the age of agentic AI: A system-theoretical approach. *ACM Transactions on Interactive Intelligent Systems.*

Cohen, Y., Biton, A., & Shoval, S. (2025). Fusion of computer vision and AI in collaborative robotics: A review and future prospects. *Applied Sciences*, *15*(14), 5678.

Dong, Y., Wang, S., Huang, Q., Berg, R. W., Li, G., & He, J. (2023). Neural decoding for intracortical brain–computer interfaces. *Cyborg and Bionic Systems*, *5*, 0075.

Edelman, B. J., Zhang, S., Schalk, G., Brunner, P., Müller-Putz, G., Guan, C., & He, B. (2024). Non-invasive brain–computer interfaces: State of the art and trends. *IEEE Reviews in Biomedical Engineering*, *17*, 1–28.

Etemad-Sajadi, R., Soussan, A., & Schöpfer, T. (2022). How ethical issues raised by human–robot interaction can impact the intention to use the robot? *International Journal of Social Robotics*, *14*(6), 1597–1611.

Finch, M., Ní Bhroin, N., & Krüger, S. (2023). Unlearning, relearning, staying with the trouble: Scenarios and the future of education. *Prospects*, *53*(1–2), 1–18.

Gerlich, M. (2023). Perceptions and acceptance of artificial intelligence: A multi-dimensional study. *Social Sciences*, *12*(9), 430.

Giordano, G., Babu, S. P. M., & Mazzolai, B. (2023). Soft robotics towards sustainable development goals and climate actions. *Frontiers in Robotics and AI*, *10*, 1122334.

Göngör, F., & Tutsoy, O. (2024). On the remarkable advancement of assistive robotics in human–robot interaction-based health-care applications: An exploratory overview of the literature. *International Journal of Human–Computer Interaction*, *40*(8), 1643–1675.

Haidegger, T., Mai, V., Mörch, C. M., Boesl, D. O., Jacobs, A., Rao, B., Khamis, A., Lach, L., & Vanderborght, B. (2023). Robotics: Enabler and inhibitor of the sustainable development goals. *Sustainable Production and Consumption*, *39*, 567–585.

Hopko, S. K., Wang, J., & Mehta, R. K. (2022). Human factors considerations and metrics in shared space human–robot collaboration: A systematic review. *Frontiers in Robotics and AI*, *9*, 829079.

Hwang, J. Y. (2024). Ethics of artificial intelligence: Examining moral accountability in autonomous decision-making systems. *World Journal of Advanced Research and Reviews*, *22*(3), 311–322.

Järvelä, S., Zhao, G., Nguyen, A., & Chen, H. (2025). Hybrid intelligence: Human–AI coevolution and learning. *British Journal of Educational Technology*, *56*(2), 321–338.

Kamalov, F., Santandreu Calonge, D., & Gurrib, I. (2023). New era of artificial intelligence in education: Towards a sustainable multifaceted revolution. *Sustainability*, *15*(10), 8234.

Karas, K., Pozzi, L., Pedrocchi, A., Braghin, F., & Roveda, L. (2023). Brain–computer interface for robot control with eye artifacts for assistive applications. *Scientific Reports*, *13*, 17421.

Kim, S., Anthis, J. R., & Sebo, S. (2024). A taxonomy of robot autonomy for human-robot interaction. In *Proceedings of the 19th ACM/IEEE International Conference on Human-Robot Interaction* (pp. 1–12).

Licardo, J. T., Domjan, M., & Orehovački, T. (2024). Intelligent robotics—A systematic review of emerging technologies and trends. *Electronics*, *13*(3), 522.

Lv, Z., Poiesi, F., Dong, Q., Lloret, J., & Song, H. (2022). Deep learning for intelligent human–computer interaction. *Applied Sciences*, *12*(22), 11674.

Mårell-Olsson, E., Bensch, S., Hellström, T., Alm, H., Hyllbrant, A., Leonardson, M., & Westberg, S. (2025). Navigating the human–robot interface—Exploring human interactions and perceptions with social and telepresence robots. *Applied Sciences*, *15*(2), 456.

Martínez Elizalde, L. B., Astengo Noguez, C., Landa Cavazos, M. R., & Salgado Garza, L. R. (2025). Bridging AI, robotics, and software engineering: An interdisciplinary approach for learning emerging technologies. In *2025 IEEE Global Engineering Education Conference (EDUCON)* (pp. 1–8). IEEE.

Palenski, T., Hills, L., Unnikrishnan, S., & Eynon, R. (2024). How AI works: Reconfiguring lifelong learning. *Postdigital Science and Education*, *6*(3), 455–478.

Puerta-Beldarrain, M., Gómez-Carmona, O., Sánchez-Corcuera, R., Casado-Mansilla, D., López-de-Ipiña, D., & Chen, L. (2024). A multifaceted vision of the human–AI collaboration: A comprehensive review. *IEEE Access*, *12*, 105233–105260.

Qu, H., Zeng, F., Tang, Y., Shi, B., Wang, Z., Chen, X., & Wang, J. (2022). The clinical effects of brain–computer interface with robot on upper-limb function for post-stroke rehabilitation: A meta-analysis and systematic review. *Disability and Rehabilitation: Assistive Technology*, *17*(8), 936–949.

Qu, J., Cui, G., Li, Z., Fang, S., Zhang, X., Liu, A., & Wang, X. (2024). Advanced flexible sensing technologies for soft robots. *Advanced Functional Materials*, *34*(25), 2315678.

Radanliev, P. (2025). AI ethics: Integrating transparency, fairness, and privacy in AI development. *Applied Artificial Intelligence*, *39*(2), 120–143.

Sarker, A., Islam, T. U., & Islam, M. R. (2024). A review on recent trends of bioinspired soft robotics: Actuators, control methods, materials selection, sensors, challenges, and future prospects. *Advanced Intelligent Systems*, *6*(10), 2400123.

Shengelia, R. (2025). Artificial intelligence and labor market dynamics: Employment problems and development trends. *Economics*, *11*(2), 55–72.

Stahl, B., & Eke, D. (2024). The ethics of ChatGPT: Exploring the ethical issues of an emerging technology. *International Journal of Information Management*, *74*, 102733.

Urrea, C., & Kern, J. (2025). Recent advances and challenges in industrial robotics: A systematic review of technological trends and emerging applications. *Processes*, *13*(3), 512.

Verma, S., & Singh, V. (2024). The employees' intention to work in artificial intelligence-based hybrid environments. *IEEE Transactions on Engineering Management*, *71*(2), 456–469.

Weidemann, C., Mandischer, N., van Kerkom, F., Corves, B., Hüsing, M., Kraus, T., & Garus, C. (2023). Literature review on recent trends and perspectives of collaborative robotics in Work 4.0. *Robotics*, *12*(3), 66.

Yin, S., Yao, D. R., Song, Y., Heng, W., Ma, X., Han, H., & Gao, W. (2024). Wearable and implantable soft robots. *Chemical Reviews*, *124*(3), 1456–1512.

Yin, Z., Li, Z., Zhai, X., & Gao, J. (2026). Human–machine mutual trust-based shared control framework for intelligent vehicles. *Advanced Intelligent Systems*, *8*(1), 2300456.

Yu, P., Xu, H., Hu, X., & Deng, C. (2023). Leveraging generative AI and large language models: A comprehensive roadmap for healthcare integration. *Healthcare*, *11*(19), 2604.

Zhao, J., & Gómez Fariñas, B. (2022). Artificial intelligence and sustainable decisions. *European Business Organization Law Review*, *23*(4), 743–772.

Zhou, S., Li, Y., Wang, Q., & Lyu, Z. (2024). Integrated actuation and sensing: Toward intelligent soft robots. *Cyborg and Bionic Systems*, *5*, 0035.

Index

Note: Page numbers in italics indicate figures.